AF539328

Enzyme and Food Biotechnology

Enzyme and Food Biotechnology

Dr. Pankaj Goyal

RANDOM PUBLICATIONS
NEW DELHI (INDIA)

Enzyme and Food Biotechnology

ISBN 978-93-5111-210-5

Published in 2014 in India by

RANDOM PUBLICATIONS

4376-A/4B, Gali Murari Lal, Ansari Road
New Delhi-110 002
Phone : +91-11-43580356, +91-11-23289044
e-mail: randomexports@gmail.com, sales@randompublications.com, info@randompublications.com

Reprinted 2021

Type Setting by : Keystoneprintads, Delhi-110051
Digitally Printed at: Replika Press Pvt. Ltd.

Preface

A communication network which does not depend on any physical connection needed to between two communication entities and have the flexibility to be mobile during communication. The current GSM and CDMA technology offers Mobile Communication. Mobile communication allows transmission of voice and multimedia data via a computer or a mobile device without having connected to any physical or fixed link. Mobile communication is evolving day by day and has become a must have for everyone. Mobile communication is the exchange of voice and data using a communication infrastructure at the same time regardless of any physical link.

Mobile Communications and Networking research is focused on next generation cellular wireless data network architectures. New technologies for improving spectral efficiency, spectrum utility and spatial reuse of spectrum on the radio-access network are being developed. On the mobile core network, research is oriented towards efficiently handling the exploding amount of mobile data traffic, enabling new revenue generating services for mobile operators and network CAPEX/OPEX reduction techniques. A variety of approaches including analysis, simulations, modeling, as well as experimental prototype design and network measurements are used to validate the technologies developed.

Mobile Technology has groomed a lot in the past few years, major reasons for rapid advancements in the mobile network technology is requirements for being mobile or connected to moves. Latest mobile handsets offer features which one had never thought off, ultimately it forces mobile network companies to bring these features in practice use to take commercial advantages.

Cellular is one of the fastest ongoing growing telecommunication industry in the world. It may easily be one of those industries which may never suffer from the economics of the world.

Wireless communications are a type of data communication that is performed and delivered wirelessly. This is a broad term that incorporates all procedures and forms of connecting and communicating between two or more devices using a wireless signal through wireless communication technologies and devices. Wireless communication is among technology's biggest contributions to mankind. Wireless communication involves the transmission of information over a distance without the help of wires, cables or any other forms of electrical conductors. The transmitted distance can be anywhere between a few meters (for example, a television's remote control) and thousands of kilometres (for example, radio communication).

The book is self-explanatory to a large extent and also serves as convenient reference for engineers working in the Mobile Communications field.

I thank all members of my team who have helped in the preparation of the book. My special thanks go to "Random Publications" who have published the book.

– Dr. Pankaj Goyal

Contents

1

Process of Enzymes Activity

Enzymes are catalysts that optimize cell activity while minimizing the amount of energy needed to achieve a specific reaction. Enzymes are also energized protein molecules found in every living cell, and are necessary for life. There are over 2000 known enzymes, each of which is involved with one specific chemical reaction. They are any of various proteins, originating from living cells and capable of producing certain chemical changes in organic substances by catalytic action, such as digestion. These proteins, and their function (s), are determined by their shape. In cells and organisms, most reactions are catalyzed by enzymes, which are regenerated during the course of a reaction. Biological catalysts are physiologically important because they speed up rates of reactions that would otherwise be too slow to support life. Our bodies naturally produce digestive and metabolic enzymes as they are needed. Specifically, the pancreas produces enzymes that break down foods into nutrients the body can use for energy and other bodily functions. The names of enzymes often include the substrate or substance on which they act, joined with an-ase ending. For example, lactase acts upon lactose and maltase acts on maltose to produce glucose. Sometimes they are named for their reaction product, for example, sucrase is often called invertase, because invertase is the result of the reaction of sucrose. Enzymes can also bear a name the describes the reaction that is catalyzed. An example of this includes the use of the name oxidase, because oxidase is involved in an oxidation reaction.

ENZYME CLASSIFICATIONS

Traditionally, enzymes were simply assigned names by the investigator who discovered the enzyme. As knowledge expanded, systems of enzyme classification became more comprehensive and complex. Currently enzymes are grouped into six functional classes by the International Union of Biochemists (I.U.B.). These rules give each enzyme a unique number. The I.U.B. system also specifies a textual name for each enzyme. The enzyme's name is comprised of the names of the substrate(s), the product(s) and the enzyme's functional class. Because many enzymes, such as alcohol dehydrogenase, are widely known in the scientific community by their common names, the change to I.U.B.-

approved nomenclature has been slow. In everyday usage, most enzymes are still called by their common name.

Number	Classification	Biochemical Properties
1	Oxidoreductases	Act on many chemical groupings to add or remove hydrogen atoms.
2	Transferases and acceptor	Transfer functional groups between donor molecules. Kinases are specialized transferases that regulate metabolism by transferring phosphate from ATP to other molecules.
3	Hydrolases	Add water across a bond, hydrolyzing it.
4.	Lyases	Add water, ammonia or carbon dioxide across double bonds, or remove these elements to produce double bonds.
5	Isomerases	Carry out many kinds of isomerization: L to D isomerizations, mutase reactions (shifts of chemical groups) and others.
6	Ligases	Catalyze reactions in which two chemical groups are joined (or ligated) with the use of energy from ATP.

Enzymes are also classified on the basis of their composition. Enzymes composed wholly of protein are known as simple enzymes in contrast to complex enzymes, which are composed of protein plus a relatively small organic molecule. Complex enzymes are also known as holoenzymes. In this terminology the protein component is known as the apoenzyme, while the non-protein component is known as the coenzyme or prosthetic group where prosthetic complex in which the small organic molecule is bound to the apoenzyme by covalent bonds; when the binding between the apoenzyme and non-protein components is non-covalent, the small organic molecule is called a coenzyme. Many prosthetic groups and coenzymes are water-soluble derivatives of vitamins. It should be noted that the main clinical symptoms of dietary vitamin insufficiency generally arise from the malfunction of enzymes, which lack sufficient cofactors derived from vitamins to maintain homeostasis. The non-protein component of an enzyme may be as simple as a metal ion or as complex as a small non-protein organic molecule. Enzymes that require a metal in their composition are known as metalloenzymes if they bind and retain their metal atom(s) under all conditions with very high affinity. Those which have a lower affinity for metal ion, but still require the metal ion for activity, are known as metal-activated enzymes.

ENZYMES AND LIFE PROCESSES

The living cell is the site of tremendous biochemical activity called metabolism. This is the process of chemical and physical change which goes on continually in the living organism. Build-up of new tissue, replacement of old tissue, conversion of food to energy, disposal of waste materials, reproduction - all the activities that we characterize as "life." This building up and tearing down takes place in the face of an apparent paradox. The greatest majority of these biochemical reactions do not take place spontaneously. The phenomenon of catalysis makes possible biochemical reactions necessary for all life processes. Catalysis is defined as the acceleration of a chemical reaction by some substance which itself undergoes no permanent chemical change. The catalysts of biochemical reactions are enzymes and are responsible for bringing about almost all of the chemical reactions in living organisms. Without enzymes, these reactions take place at a rate far too slow for the pace of metabolism. The oxidation of a fatty acid to carbon dioxide and water is not a gentle process in a test tube - extremes of pH, high temperatures and corrosive chemicals are required. Yet in the body, such a reaction takes place smoothly and rapidly within a narrow range of pH and temperature. In the laboratory, the average protein must be boiled for about 24 hours in a 20% HCl solution to achieve a complete breakdown. In the body, the breakdown takes place in four hours or less under conditions of mild physiological temperature and pH.

CHEMICAL NATURE OF ENZYMES

All known enzymes are proteins. They are high molecular weight compounds made up principally of chains of amino acids linked together by peptide bonds.

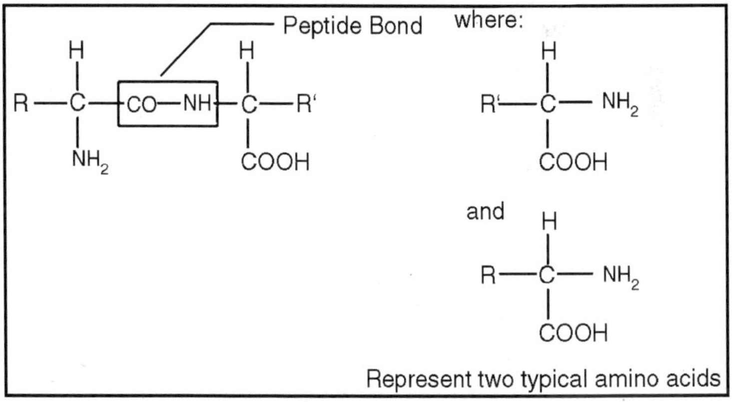

Peptide Bond

$R-CH(NH_2)-[CO-NH]-CH(COOH)-R'$

where:

$R'-CH(COOH)-NH_2$

and

$R-CH(COOH)-NH_2$

Represent two typical amino acids

Fig. Typical Protein Structure - Two Amino Acids Joined by a Peptide Bond.

Enzymes can be denatured and precipitated with salts, solvents and other reagents. They have molecular weights ranging from 10,000 to 2,000,000. Many enzymes require the presence of other compounds - cofactors - before their catalytic activity can be exerted. This entire active complex is referred to as the holoenzyme; *i.e.*, apoenzyme (protein portion) plus the cofactor (coenzyme, prosthetic group or metal-ion-activator) is called the holoenzyme.

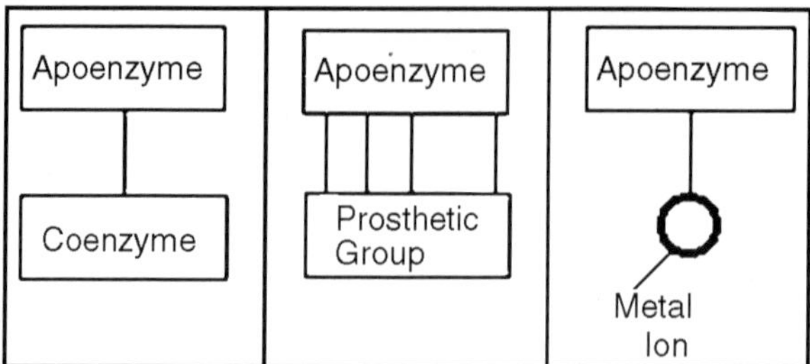

Fig. Holoenzymes- Apoenzymes Plus Various Types of Cofactors.

Apoenzyme + Cofactor = Holoenzyme

According to Holum, the cofactor may be:

- *A coenzyme*: A non-protein organic substance which is dialyzable, thermostable and loosely attached to the protein part.
- *A prosthetic group*: An organic substance which is dialyzable and thermostable which is firmly attached to the protein or apoenzyme portion.
- *A metal-ion-activator*: These include K^+, Fe^{++}, Fe^{+++}, Cu^{++}, Co^{++}, Zn^{++}, Mn^{++}, Mg^{++}, Ca^{++}, and Mo^{+++}.

Specificity of Enzymes

One of the properties of enzymes that makes them so important as diagnostic and research tools is the specificity they exhibit relative to the reactions they catalyze. A few enzymes exhibit absolute specificity; that is, they will catalyze only one particular reaction. Other enzymes will be specific for a particular type of chemical bond or functional group.

In general, there are four distinct types of specificity:

1. *Absolute specificity*: The enzyme will catalyze only one reaction.
2. *Group specificity*: The enzyme will act only on molecules that have specific functional groups, such as amino, phosphate and methyl groups.
3. *Linkage specificity*: The enzyme will act on a particular type of chemical bond regardless of the rest of the molecular structure.
4. *Stereochemical specificity*: The enzyme will act on a particular steric or optical isomer.

Though enzymes exhibit great degrees of specificity, cofactors may serve many apoenzymes. For example, nicotinamide adenine dinucleotide (NAD) is a coenzyme for a great number of dehydrogenase reactions in which it acts as a hydrogen acceptor. Among them are the alcohol dehydrogenase, malate dehydrogenase and lactate dehydrogenase reactions.

Naming and Classification

Except for some of the original enzymes such as pepsin, rennin, and trypsin, most enzyme names end in "ase". The International Union of Biochemistry (I.U.B.) initiated standards of enzyme nomenclature which recommend that enzyme names indicate both the substrate acted upon and the type of reaction catalyzed. Under this system, the enzyme uricase is called urate: O_2 oxidoreductase, while the enzyme glutamic oxaloacetic transaminase (GOT) is called L-aspartate: 2-oxoglutarate aminotransferase.

Enzymes can be classified by the kind of chemical reaction catalyzed:

- Addition or removal of water
  - *Hydrolases*: These include esterases, carbohydrases, nucleases, deaminases, amidases, and proteases
  - Hydrases such as fumarase, enolase, aconitase and carbonic anhydrase
- Transfer of electrons
  - Oxidases
  - Dehydrogenases
- Transfer of a radical
  - *Transglycosidases*: Of monosaccharides
  - *Transphosphorylases and phosphomutases*: Of a phosphate group
  - *Transaminases*: Of amino group
  - *Transmethylases*: Of a methyl group
  - *Transacetylases*: Of an acetyl group
- Splitting or forming a C-C bond
  - Desmolases
- Changing geometry or structure of a molecule
  - Isomerases
- Joining two molecules through hydrolysis of pyrophosphate bond in ATP or other triphosphate
  - Ligases

TYPES OF ENZYMES

- Simple enzymes are those composed completely of proteins
- Complex enzymes are those composed of protein and a small organic molecule (s) (also known as holoenzymes)

Enzymes are Categorized by their Function (s)**:**

- Metabolic enzymes catalyze and regulate every biochemical reaction that occurs in your body and are essential to cellular function and health.
- Digestive enzymes are secreted along the digestive tract, break down food into nutrients and waste, and allow the nutrients found in foods to be absorbed into the bloodstream. The pancreas produces most digestive enzymes, but the liver, gallbladder, small intestines, stomach and colon also play critical roles in the production of these enzymes.

  Examples of digestive enzymes: lipase, protease, amylase, ptyalin, pepsin and trypsin.

- Food enzymes enter the body through the consumption of raw foods or enzyme supplements. Cooking and processing destroys most enzymes in food.
- Plant Based enzymes are synthetically grown, help break down fat, protein and carbohydrates and function within a broad pH range. Plant enzymes, commonly bromelin (pineapple) and papain (papaya) play an important role in more complete digestion of all foods. They are activated at a temperature higher than normal body temperature, also making them a good anti-inflammatory.
- Proteoplytic enzymes are enzymes that digest proteins, including Trypsin, Chymotrypsin, Pancreatin, Bromelin and Papain. Some are produced by the pancreas and others are supplemental from animals or plants. The primary use of proteolytic enzymes in supplements is, are as digestive aids. They are also used in drugs as anti-inflammatory agents and pain relievers.

WORKING PROCESS OF ENZYMES

TYPES OF ENZYMATIC REACTIONS

When looking at enzyme names such as 'transketolase' or 'phosphorylase', you will note that these names don't tell you exactly what reactions the enzymes may catalyze. A complete description should mention the coenzymes required, the substrates and the particular bonds in the substrates that are being severed or created.

A nomenclature that meets these criteria has been developed by the Enzyme Commission of the IUBMB (International Union of Biochemistry and Molecular Biology). In the IUBMB nomenclature, the enzyme transketolase bears the formidable name:

Sedoheptulose-7-phosphate: D-glyceraldehyde-3-phosphate glycolaldehydetransferase. Such names, of course, are rather lengthy, and their use is not very widespread. To make the tasks of tracking and bookkeeping more manageable, these names are supplemented with numeric codes. In the IUBMB scheme, enzymes are put into one out of six classes according to the reactions they catalyze.

These classes are:

- *Oxidoreductases*. These catalyze redox reactions, frequently involving one of the coenzymes NAD^+, $NADP^+$, or FAD.
- *Transferases*. These bring about the transfer of functional groups – e.g., phosphate groups from ATP to another metabolite, which activates the latter and sets it up for subsequent reaction steps.
- *Hydrolases*. These catalyze hydrolysis reactions – e.g., such as those involved in digestion of foodstuffs.
- *Lyases* – these effect elimination reactions that result in the formation of double bonds.

- *Isomerases*. These facilitate the interconversion of isomers. We will meet two examples as soon as we get into glycolysis.
- Ligases, which form new covalent bonds at the expense of ATP hydrolysis.

Of course, within each of these main classes, there are subclasses and sub-sub classes that correspond to details of substrates and mechanisms of the enzyme reactions. Each individual enzyme activity is assigned an individual number within a sub-sub class, so that we wind up with a four-figure designation, which is preceded by the letters 'EC' (Enzyme Commission). One good thing about this classification is that it is rarely used – the 'recommended names', which most of the time happen to be the traditional ones, are used instead. The other good thing is that it has a sound appreciation of priorities. The single most important enzyme in student lifestyle – namely, alcohol dehydrogenase (or, as IUBMB puts it, alcohol:NAD oxidoreductase). This beneficial enzyme, residing in the liver, degrades ethanol, and without it you would be drunk all the time!

Enzyme Kinetics: Basic Enzyme Reactions

Enzymes are catalysts and increase the speed of a chemical reaction without themselves undergoing any permanent chemical change. They are neither used up in the reaction nor do they appear as reaction products.

The basic enzymatic reaction can be represented as follows where E represents the enzyme catalyzing the reaction, S the substrate, the substance being changed, and P the product of the reaction.

Enzyme Kinetics: Energy Levels

Chemists have known for almost a century that for most chemical reactions to proceed, some form of energy is needed. They have termed this quantity of energy, "the energy of activation." It is the magnitude of the activation energy which determines just how fast the reaction will proceed. It is believed that enzymes lower the activation energy for the reaction they are catalyzing. Figure illustrates this concept.

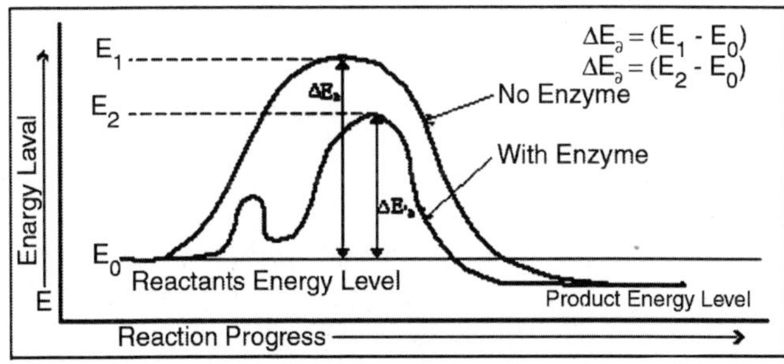

Fig. Energy of Activation Δ_∂ E is less than Δ_∂ E due to the Effect of the Enzyme on the Substrate.

The enzyme is thought to reduce the "path" of the reaction. This shortened path would require less energy for each molecule of substrate converted to product. Given a total amount of available energy, more molecules of substrate would be converted when the enzyme is present (the shortened "path") than when it is absent. Hence, the reaction is said to go faster in a given period of time.

ENZYME KINETICS

A theory to explain the catalytic action of enzymes was proposed by the Swedish chemist Savante Arrhenius in 1888. He proposed that the substrate and enzyme formed some intermediate substance which is known as the enzyme substrate complex.

The reaction can be represented as:

$$\underset{\text{substrate}}{S} + \underset{\text{enzyme}}{E} \longrightarrow \underset{\text{enzyme substrate complex}}{ES}$$

If this reaction is combined with the original reaction equation, the following results:

$$\underset{\text{substrate}}{S} + \underset{\text{enzyme}}{E} \longrightarrow \underset{\text{enzyme substrate complex}}{ES} \longrightarrow \underset{\text{Product}}{P} + \underset{\text{enzyme}}{E}$$

The existence of an intermediate enzyme-substrate complex has been demonstrated in the laboratory, for example, using catalase and a hydrogen peroxide derivative. At Yale University, Kurt G. Stern observed spectral shifts in catalase as the reaction it catalyzed proceeded. This experimental evidence indicates that the enzyme first unites in some way with the substrate and then returns to its original form after the reaction is concluded.

Chemical Equilibrium

The study of a large number of chemical reactions reveals that most do not go to true completion. This is likewise true of enzymatically-catalyzed reactions.

This is due to the reversibility of most reactions. In general:

$$A + B \xrightarrow{K+1} C + D \quad \text{forward reaction}$$

$$C + D \xrightarrow{K-1} A + B \quad \text{reverse reaction}$$

where K^{+1} is the forward reaction rate constant and K^{-1} is the rate constant for the reverse reaction.

Combining the two reactions gives:

$$A + B \underset{K-1}{\overset{K+1}{\rightleftarrows}} C + D$$

Applying this general relationship to enzymatic reactions allows the equation:

$$E + S \underset{K-1}{\overset{K+1}{\rightleftharpoons}} ES \underset{K-2}{\overset{K+2}{\rightleftharpoons}} p + E$$

Equilbrium, a steady state condition, is reached when the forward reaction rates equal the backward rates. This is the basic equation upon which most enzyme activity studies are based.

Factors Affecting Enzyme Activity

Knowledge of basic enzyme kinetic theory is important in enzyme analysis in order both to understand the basic enzymatic mechanism and to select a method for enzyme analysis. The conditions selected to measure the activity of an enzyme would not be the same as those selected to measure the concentration of its substrate. Several factors affect the rate at which enzymatic reactions proceed-temperature, pH, enzyme concentration, substrate concentration, and the presence of any inhibitors or activators.

Enzyme Concentration

In order to study the effect of increasing the enzyme concentration upon the reaction rate,the substrate must be present in an excess amount; i.e., the reaction must be independent of the substrate concentration. Any change in the amount of product formed over a specified period of time will be dependent upon the level of enzyme present. These reactions are said to be "zero order" because the rates are independent of substrate concentration, and are equal to some constant k. The formation of product proceeds at a rate which is linear with time. The addition of more substrate does

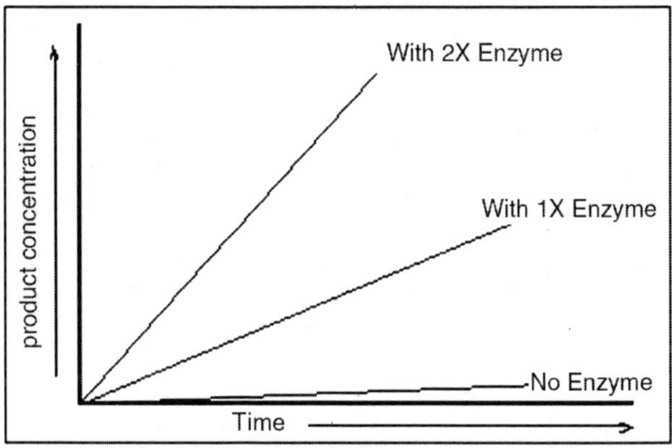

Fig. "Zer order" Reaction Rate is Independent of Substrate Concentration.

not serve to increase the rate. In zero order kinetics, allowing the assay to run for double time results in double the amount of product.

Table. Reaction Orders with Respect to Substrate Concentration

Order	Rate Equation	Comments
Zero	rate = k	rate is independent of substrate concentration
First	rate = k[S]	rate is proportional to the first power of substrate concentration
Second	rate = $k[S][S]=k[S]^2$	rate is proportional to the square of the substrate concentration
Second	rate = $k[S_1][S_2]$	rate is proportional to the first power of each of two reactants

The amount of enzyme present in a reaction is measured by the activity it catalyzes. The relationship between activity and concentration is affected by many factors such as temperature, pH, etc. An enzyme assay must be designed so that the observed activity is proportional to the amount of enzyme present in order that the enzyme concentration is the only limiting factor.

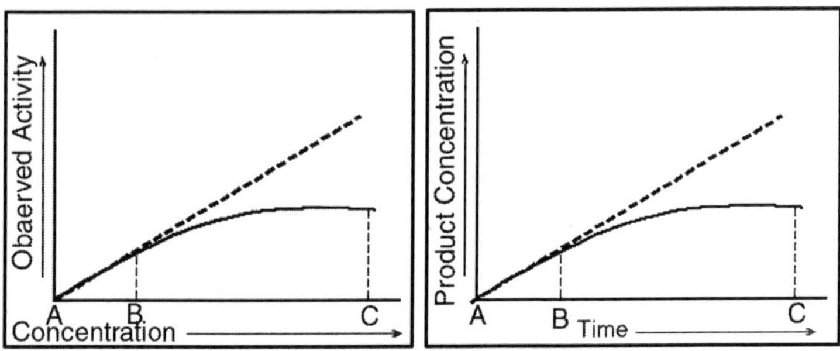

Fig. (a) Activity Concentration. (b) Reaction rate limited by substrate concentration

It is satisfied only when the reaction is zero order. In Figure, activity is directly proportional to concentration in the area AB, but not in BC. Enzyme activity is generally greatest when substrate concentration is unlimiting. When the concentration of the product of an enzymatic Reaction is plotted against time, a similar curve results. Between A and B, the curve represents a zero order reaction; that is, one in which the rate is constant with time. As substrate is used up, the enzyme's active sites are no longer saturated, substrate concentration becomes rate limiting, and the reaction becomes first order between B and C. To measure enzyme activity ideally, the measurements must be made in that portion of the curve where the reaction is zero order.

A reaction is most likely to be zero order initially since substrate concentration is then highest. To be certain that a reaction is zero order, multiple measurements of product (or substrate) concentration must be made. Three types of reactions which might be encountered in enzyme assays and shows the problems which might be enountered if only single measurements are made.

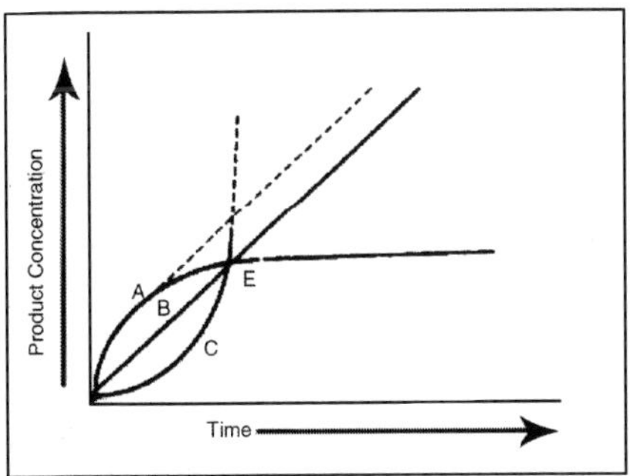

Fig. Leading, Lagging and Liner Reactions

B is a straight line representing a zero order reaction which permits accurate determination of enzyme activity for part or all of the reaction time. This reaction is zero order initially and then slows, presumably due to substrate exhaustion or product inhibition. This type of reaction is sometimes referred to as a "leading" reaction. True "potential" activity is represented by the dotted line. Curve C represents a reaction with an initial "lag" phase. Again the dotted line represents the potentially measurable activity. Multiple determinations of product concentration enable each curve to be plotted and true activity determined. A single end point determination at E would lead to the false conclusion that all three samples had identical enzyme concentration.

Substrate Concentration

It has been shown experimentally that if the amount of the enzyme is kept constant and the substrate concentration is then gradually increased, the reaction velocity will increase until it reaches a maximum. After this point, increases in substrate concentration will not increase the velocity (Δ A/ Δ).

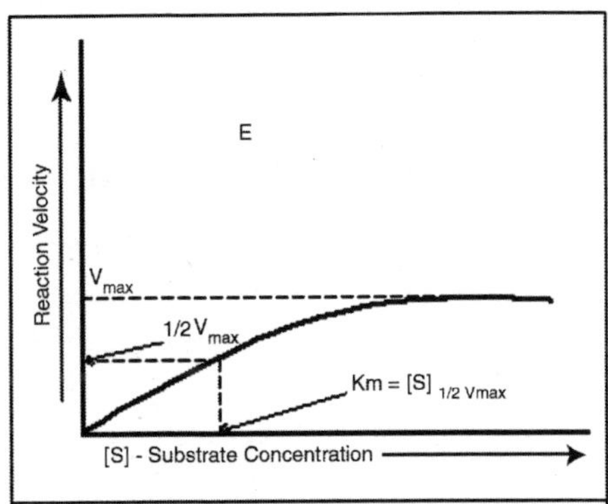

Fig. Effect of SConcentration

It is theorized that when this maximum velocity had been reached, all of the available enzyme has been converted to ES, the enzyme substrate complex.

This point on the graph is designated V_{max}. Using this maximum velocity and equation, Michaelis developed a set of mathematical expressions to calculate enzyme activity in terms of reaction speed from measurable laboratory data.

$$E + S \underset{K-1}{\overset{K+1}{\rightleftarrows}} ES \underset{K-2}{\overset{K+2}{\rightleftarrows}} p + E$$

The Michaelis constant Km is defined as the substrate concentration at 1/2 the maximum velocity. Using this constant and the fact that Km can also be defined as:

$$K_m = K_{-1} + K_{2/}K_{+1}$$

K^{+1}, K^{-1} and K^{+2} being the rate constants from equation. Michaelis developed the following:

$$V_1 \frac{V_{mx}[S]}{K_m + [S]}$$

V_1 = The velocity at any time

[S] = The substrate concentration at this time

V_{mx} = The highest under this set of experimental condition (pH, temperature, etc.)

K_M = The Michaselis constant for the particular enzyme being investigated Michaelis constants have been determined for many of the commonly used enzymes. The size of Km tells us several things about a particular enzyme.

- A small Km indicates that the enzyme requires only a small amount of substrate to become saturated. Hence, the maximum velocity is reached at relatively low substrate concentrations.
- A large Km indicates the need for high substrate concentrations to achieve maximum reaction velocity.
- The substrate with the lowest Km upon which the enzyme acts as a catalyst is frequently assumed to be enzyme's natural substrate, though this is not true for all enzymes.

ACTIVE SITES AND CATALYTIC MECHANISMS

Open any textbook of biochemistry and you will be presented with an overwhelming number of figures depicting the crystal structures of a multitude of enzymes. Of course, the three-dimensional structures of enzymes are crucial to their functions.

Many enzymes are just regular protein molecules, composed of nothing else than the 20 standard amino acids. If you mix of all these amino acids in free form at, say, 10 mM each, this mixture will not have any significant catalytic activity. It therefore is the precise arrangement of the amino acids in the enzyme molecule that brings about the function. As the á-amino and á-carboxyl groups of the amino acids are hooked up to each other in the polypeptide chain, they usually do not directly contribute to the catalytic effect of the enzyme. Instead, it is the side chains that are directly engaged win the reaction. A very good example of this is chymotrypsin. Chymotrypsin is one of the major proteases in the human digestive tract, where its job is to knock down large protein molecules into small peptides that are then further processed by peptidases.

How does chymotrypsin do that? The enzyme-catalyzed reaction is similar to alkaline hydrolysis. In alkaline hydrolysis, a hydroxide ion, which is a strong nucleophile, attacks the carbon in the peptide bond that carries a partial positive charge. In the enzyme-catalyzed reaction, a deprotonated serine residue of the enzyme plays a role similar to that of the hydroxide ion.

Now, you know that serine normally is a neutral amino acid – its OH-group does not spontaneously dissociate, no more than the-OH group of alcohol does. The question therefore is, how does the enzyme deprotonate its own serine side chain? This is brought about by placing the serine next to a histidine and an aspartate inside the active centre. Aspartate deprotonates histidine, which in turn deprotonates the serine residue. This motive – asp, his, ser – is very widespread among proteases and esterases, so much so that it is commonly called 'the catalytic triad'. E.g., the protease trypsin and several lipases that occur in human metabolism have this motif and share the same mechanism of catalysis.

Structure and Mechanism of Chymotrypsin:

Fig. The Peptide Bond cleaved by chymotrypsin.

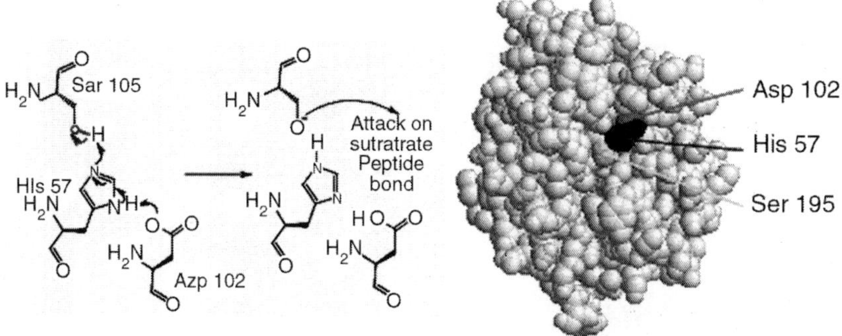

Fig. The Interplay of the Side Chains of Histidine 57, Aspartate 102, and Serine 195 that Results in the Deprotonation of Serine.

All three side chains are close to each other in the active site of the enzyme. While with many enzymes the protein molecule and its amino acid side chains are sufficient for catalysis, many others require co-enzymes for their catalytic activity. Very often, both active-site amino acid side chains and coenzymes are required. For example, amino acid transaminases have a molecule of the coenzyme pyridoxalphosphate bound to the active site, which co-operates with an active site lysine in the enzyme's catalytic activity. Most enzymes have just one active site, or if they are multimeric one active site per subunit. However, there are exceptions: Fatty acid synthase has as many as eight different active sites on each subunit. Multi-enzyme complexes such as pyruvate dehydrogenase have one active site per subunit but combine different types of subunits and enzyme activities in one functional assembly.

Fig. The Enzyme-Catalized reaction.

Fig. The Base-Catalyzed Reaction for Comparison.

EFFECTS OF INHIBITORS ON ENZYME ACTIVITY

Enzyme inhibitors are substances which alter the catalytic action of the enzyme and consequently slow down, or in some cases, stop catalysis. There are three common types of enzyme inhibition-competitive, non-competitive and substrate inhibition.

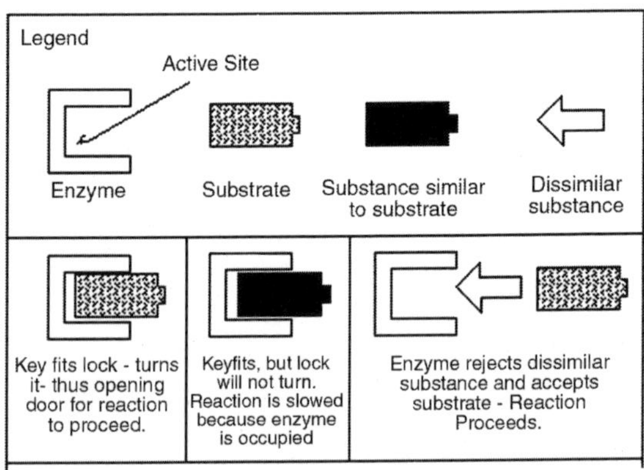

Fig. Lock-key Theory-Competitive Inhibition

Most theories concerning inhibition mechanisms are based on the existence of the enzyme-substrate complex ES. As mentioned earlier, the existence of temporary ES structures has been verified in the laboratory. Competitive inhibition occurs when the substrate and a substance resembling the substrate are both added to the enzyme. A theory called the "lock-key theory" of enzyme catalysts can be used to explain why inhibition occurs.

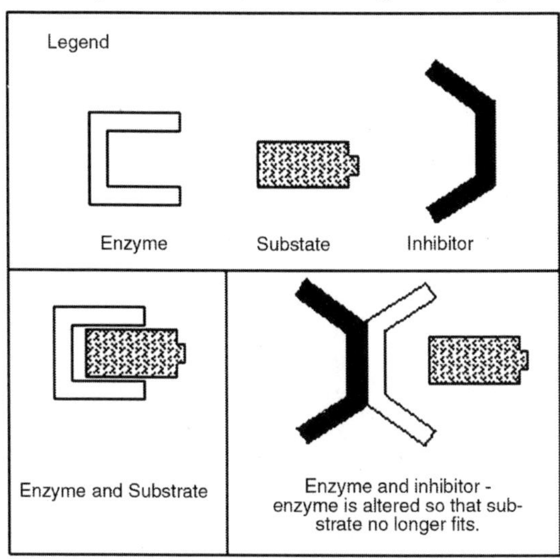

Fig. Non-Competitive Inhibition

The lock and key theory utilizes the concept of an "active site." The concept holds that one particular portion of the enzyme surface has a strong affinity for the substrate. The substrate is held in such a way that its conversion to the reaction products is more favorable. If we consider the enzyme as the lock and the substrate the key-the key is inserted in the lock, is turned, and the door is opened and the reaction proceeds.

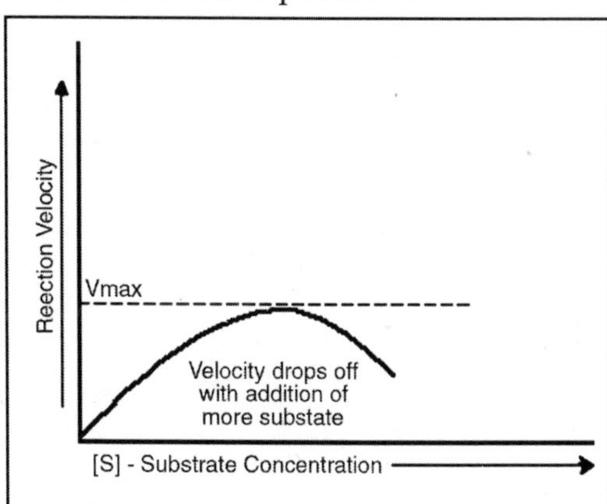

Fig. Substrate Become Rate-inhibiting

However, when an inhibitor which resembles the substrate is present, it will compete with the substrate for the position in the enzyme lock. When the inhibitor wins, it gains the lock position but is unable to open the lock. Hence, the observed reaction is slowed down because some of the available enzyme sites are occupied by the inhibitor. If a dissimilar substance which does not fit the site is present, the enzyme rejects it, accepts the substrate, and the reaction proceeds normally. Non-competitive inhibitors are considered to be substances which when added to the enzyme alter the enzyme in a way that it cannot accept the substrate. Substrate inhibition will sometimes occur when excessive amounts of substrate are present. Figure shows the reaction velocity decreasing after the maximum velocity has been reached. Additional amounts of substrate

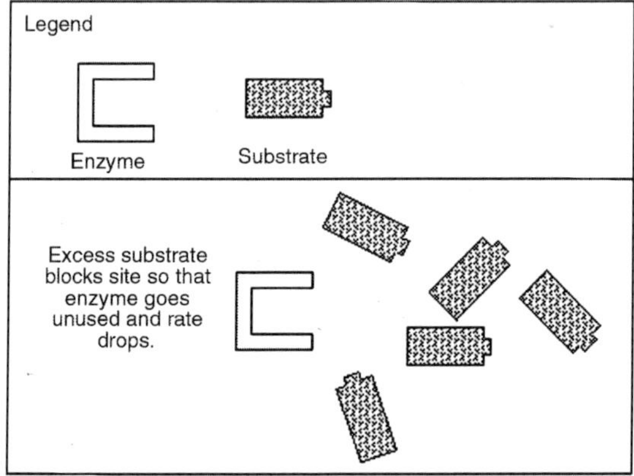

Fig. Substrate Inhibition

added to the reaction mixture after this point actually decrease the reaction rate. This is thought to be due to the fact that there are so many substrate molecules competing for the active sites on the enzyme surfaces that they block the sites and prevent any other substrate molecules from occupying them. This causes the reaction rate to drop since all of the enzyme present is not being used.

Temperature Effects

Like most chemical reactions, the rate of an enzyme-catalyzed reaction increases as the temperature is raised. A ten degree Centigrade rise in temperature will increase the activity of most enzymes by 50 to 100%. Variations in reaction temperature as small as 1 or 2 degrees may introduce changes of 10 to 20% in the results. In the case of enzymatic reactions, this is complicated by the fact that many enzymes are adversely affected by high temperatures. The reaction rate increases with temperature to a maximum level, then abruptly declines with further increase of temperature. Because most animal enzymes rapidly become denatured at temperatures above 40°C, most enzyme determinations are carried out somewhat below that temperature. Over a period of time, enzymes will be deactivated at even moderate temperatures. Storage of enzymes at 5°C or below is generally the most suitable. Some enzymes lose their activity when frozen.

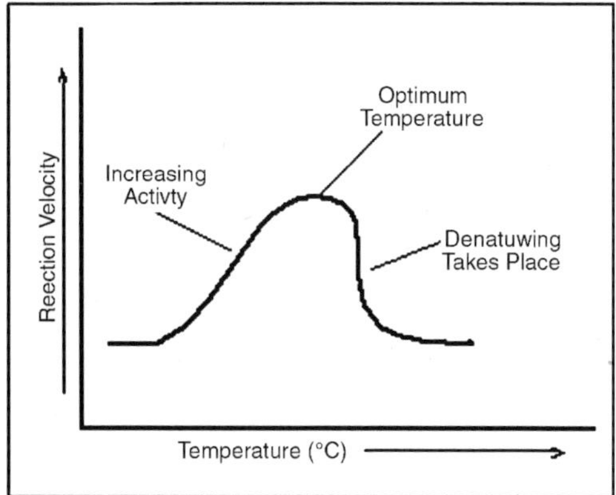

Fig. Effect of Temperature on Reaction Rate

Effects of pH

Enzymes are affected by changes in pH. The most favorable pH value-the point where the enzyme is most active-is known as the optimum pH.

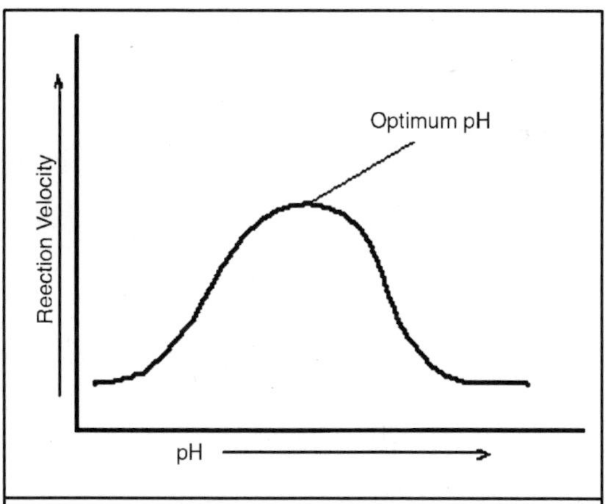

Extremely high or low pH values generally result in complete loss of activity for most enzymes. pH is also a factor in the stability of enzymes. As with activity, for each enzyme there is also a region of pH optimal stability. The optimum pH value will vary greatly from one enzyme to another, table shows: In addition to temperature and pH there are other factors, such as ionic strength, which can affect the enzymatic reaction. Each of these physical and chemical parameters must be considered and optimized in order for an enzymatic reaction to be accurate and reproducible.

Table. pH for Optimum Activity.

Enzyme	pH Optimum
Lipase (pancreas)	8.0
Lipase (stomach)	4.0-5.0
Lipase (castor oil)	4.7
Pepsin	1.5-1.6
Trypsin	7.8-8.7
Urease	7.0
Invertase	4.5
Maltase	6.1-6.8
Amylase (pancreas)	6.7-7.0
Amylase (malt)	4.6-5.2
Catalase	7.0

The term enzyme kinetics implies a study of the speed, rate or velocity of an enzyme catalysed reaction, and of the various factors which may affect this. At the heart of any study of enzyme kinetics is a knowledge of the way in which reaction velocity is altered by changes in the concentration of the enzyme's substrate and of the simple mathematics underlying this. To ease ourselves gently into this we will assume that the enzyme that we are discussing has no special features, such as allosteric properties, and catalyses the conversion of just one substrate to one product. This may seem unrelated to real enzymes, very few have just one substrate after all, but it will provide a basis which we can expand on later when we study more complex systems.

Enzyme kinetics is the study of the chemical reactions that are catalysed by enzymes, with a focus on their reaction rates. The study of an enzyme's kinetics reveals the catalytic mechanism of this enzyme, its role in metabolism, how its activity is controlled, and how a drug or a poison might inhibit the enzyme. Enzymes are usually protein molecules that manipulate other molecules — the enzymes' substrates. These target molecules bind to an enzyme's active site and are transformed into products through a series of steps known as the enzymatic mechanism. These mechanisms can be divided into single-substrate and multiple-substrate mechanisms. Kinetic studies on enzymes that only bind one substrate, such as triosephosphate isomerase, aim to measure the affinity with which the enzyme binds this substrate and the turnover rate.

When enzymes bind multiple substrates, such as dihydrofolate reductase (shown right), enzyme kinetics can also show the sequence in which these substrates bind and the sequence in which products are released. An example of enzymes that bind a single substrate and release multiple products are proteases, which cleave one protein substrate into two polypeptide products. Others join two substrates together, such as DNA polymerase linking a nucleotide to DNA. Although these mechanisms are often a complex series of steps, there is typically one rate-determining step that determines the overall kinetics. This rate-determining step may be a chemical reaction or a conformational change of the enzyme or substrates, such as those involved in the release of product(s) from the enzyme.

Knowledge of the enzyme's structure is helpful in interpreting the kinetic data. For example, the structure can suggest how substrates and products bind during catalysis; what changes occur during the reaction; and even the role of particular amino acid residues in the mechanism. Some enzymes change shape significantly during the mechanism; in such cases, it is helpful to determine the enzyme structure with and without bound substrate analogs that do not undergo the enzymatic reaction. Not all biological catalysts are protein enzymes; RNA-based catalysts such as ribozymes and ribosomes are essential to many cellular functions, such as RNA splicing and translation. The main difference between ribozymes and enzymes is that the RNA catalysts perform a more limited set of reactions, although their reaction mechanisms and kinetics can be analysed and classified by the same methods.

The reaction catalysed by an enzyme uses exactly the same reactants and produces exactly the same products as the uncatalysed reaction. Like other catalysts, enzymes do not alter the position of equilibrium between substrates and products. However, unlike normal chemical reactions, enzymes are saturable. This means as more substrate is added, the reaction rate will increase, because more active sites become occupied. This can continue until all the enzyme becomes saturated with substrate and the rate reaches a maximum. The two most important kinetic properties of an enzyme are how quickly the enzyme becomes saturated with a particular substrate, and the maximum rate it can achieve. Knowing these properties suggests what an enzyme might do in the cell and can show how the enzyme will respond to changes in these conditions.

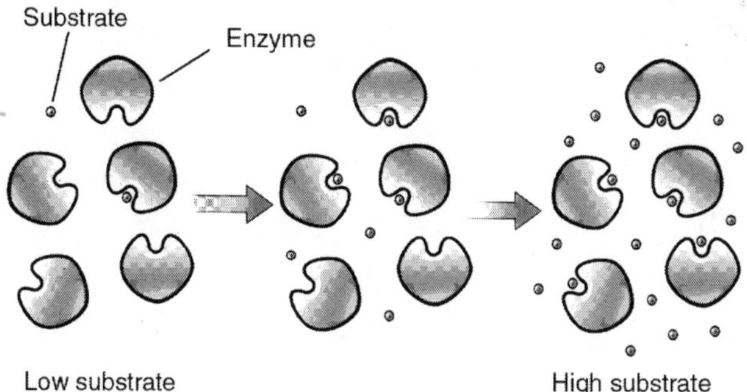

Fig. Reaction rates Increase as Substrate Concentration Increase, but Become Saturated at very High Concentrations of Substrate.

Enzyme assays are laboratory procedures that measure the rate of enzyme reactions. Because enzymes are not consumed by the reactions they catalyse, enzyme assays usually follow changes in the concentration of either substrates or products to measure the rate of reaction.

There are many methods of measurement. Spectrophotometric assays observe change in the absorbance of light between products and reactants; radiometric assays involve the incorporation or release of radioactivity to measure the amount of product made over time. Spectrophotometric assays are most convenient since they allow the rate of the reaction to be measured continuously.

Although radiometric assays require the removal and counting of samples (i. e., they are discontinuous assays) they are usually extremely sensitive and can measure very low levels of enzyme activity. An analogous approach is to use mass spectrometry to monitor the incorporation or release of stable isotopes as substrate is converted into product. The most sensitive enzyme assays use lasers focused through a microscope to observe changes in single enzyme molecules as they catalyse their reactions.

These measurements either use changes in the fluorescence of cofactors during an enzyme's reaction mechanism, or of fluorescent dyes added onto specific sites of the protein to report movements that occur during catalysis. These studies are providing a new view of the kinetics and dynamics of single enzymes, as opposed to traditional enzyme kinetics, which observes the average behaviour of populations of millions of enzyme molecules.

On the left is shown a typical progress curve for an enzyme assay. The enzyme produces product at a linear initial rate at the start of the reaction. Later in this progress curve, the rate slows down as substrate is used up or products accumulate. The length of the initial rate period depends on the assay conditions and can range from milliseconds to hours. Enzyme assays are usually set up to produce an initial rate lasting over a minute, to make measurements easier. However, equipment for rapidly mixing liquids allows fast kinetic measurements on initial rates of less than one second. These very rapid assays are essential for measuring pre-steady-state kinetics.

Most enzyme kinetics studies concentrate on this initial, linear part of enzyme reactions. However, it is also possible to measure the complete reaction curve and fit this data to a non-linear rate equation. This way of measuring enzyme reactions is called progress-curve analysis. This approach is useful as an alternative to rapid kinetics when the initial rate is too fast to measure accurately.

NATURE OF ENZYMES

Enzymes are one of the most interesting and important substances found in nature. Firts, it's important to realise that enzymes are not living things. They are inanimate-like minerals. But unlike minerals, they are made by living cells. If we were to look inside a cell we would see many different activities going on. There would be some molecules joining together and others breaking apart. These activities keep the cell alive and enzymes make these activities possible. That's why every cell of every living creature on Earth produces enzymes. In a broad sense, there are two types of enzyme. Some that help join specific molecules together to form new molecules. Others that help break specific molecules apart into separate molecules.

Enzymes play many important roles ouside the cell as well. One of the best examples of this is the digestive system. For instance, it is enzymes in your digestive system that break food down in your digestive system break food down into small molecules that can be absorbed by the body. Some enzymes in your digestive system break down starch, some proteins and others break down fats.

Four Things to Remember About Enzymes

- *Enzymes are specific*: An enzyme that is able to break fat down would not be able to dissolve protein or starch. Enzymes perform only one specific job. That means an enzyme can do its job with very few side effects. It also explains why there are so many different types of enzyme. To date, 3,000 different types have been identified and there are many more waiting to be discovered.
- *Enzymes are catalysts*: While it is true that an enzyme can only perform one speific job, it is important to know that one enzyme can perform that same job over and over again, millions of times, without being consumed in the process. And enzymes do their job best in the mild ph and temperature conditions found in nature.
- *Enzymes are efficient*: Not only do enzymes work hard, they also work with blinding speed. For instance, there is an enzyme in the liver that helps hydrogern peroxide break down into water and oxygen. What's amazing is that one enzyme can process 5 million hydrogen peroxide molecules in one minute.
- *Enzymes are natural*: Enzymes are proteins. Like all other proteins, enzymes are organic. Once they have done their job, enzymes break down swiftly and can be absorbed back into nature.

ENZYMES IN INDUSTRY

Many people believe enzyme technology is fairly new. However, that's not the case. Enzymes have been used by man from the dawn of civilization. As long as people have been eating bread and cheese, and drinking wine and beer-they have been using enzymes. That's because enzymes help to make these products.

Cheese

Take cheese for example. Most likely, cheese was discovered by ancient hunters who stored milk in the stomachs of slaughtered calves. When exposed to heat, the milk inside the container would turn into a solid-cheese. The reason for this is simple. Calves have an enzyme called chymosin in their stomachs. At ambient temperature chymosin will break down milk protein, causing milk to separate into curd and whey. If you strain away the watery whey you are left with a soft tangy curd or cheese. Chymosin derived from calf stomachs has been used in cheese making ever since.

Detergent

In more recent times, other commercial uses for enzymes have been found. Laundry detergents are a good example. As mentioned, there are certain enzymes which dissolve proteins and others which dissolve fat. Proteins and fat make up two of the major causes of stains on cloting. Grass, blood and egg are all protein stains. Lipstick, frying oil, butter, sauces, and tough stains on cuffs and collars are fat stains. to remove these stains without enzymes is difficult and requires a lot of washing at high temperatures with a lot of detergent.

Enzymes can remove these stains better, faster and in a way that is a lot kinder to the clothing and the environment. The enzymes in detergent dissolve stains in the same natural way they help digest food in your stomach-specifically and efficintly. Just a tiny amount of enzymes can dissolve stains that would require much larger amounts of detergent alone. Because of this they can reduce the amount of detergent actually found in laundry detergent. And since enzymes work best in mild conditions, washing machines using enzymatic detergants can be set at lower tempreatures, reducing electricity consumption by as much as a third. In the future, enzymes could replace more and more of the harsh chemicals found in detergents.

Many other Industries

These two examples represent just the tip of the iceberg. Enzymes are contributing to industry in many other areas. They are starting to replace petroleum-based solvents used to make vegetable seed oil; replacing harsh acids used in the production of glucose products such as corn syrup; taking over part of the role played by clorine in the paper industry; and replacing sulphides used in the tanning industry. Enzymes are also being used instead of pumice stones in the stonewashing of jeans.

Enzymes help Industry and Nature

Today enzymes are helping industry make the products used by society in a way that is less harmful to the environment. In the cases mentioned above, enzymatic processes are, in general, safer and more environment-friendly than the traditional processes they replace. That is why enzymes in particular and biotechnology in general hold such promise.

How Enzymes are Produced

Looking back to the cheese example above, chymosin is an enzyme used to turn milk into cheese. And chymosin is only found in the stomachs of young calves (as well as young sheep, goats, and a few other farm animals). Up until the 60s all the world's cheese was made from chymosin extracted from the stomachs of slaughtered calves. Then two things happened. The demand for cheese increased and the demand for alf meat decreased. Soon there were not enough slaughtered calf stomachs to produce enough high-quality chymosin required by the expanding cheese industry. To solve the chymosin shortage, it would have been very wasteful and very expensive to raise and slaughter calves just to extract a relatively small amount of chymosin from their stomachs. The cheese industry and the enzyme researchers began to look for another type of organism that produced chymosin.

Microorganisms are Natural Enzyme Factories

The researchers wanted to find an organism that was easier and less expensive to raise than a calf. An organism that reproduced quickly and wouldn't require a lot of space and food. So they began looking among the smallest and simplest organisms they knew: micoorganisms. Microorganisms are very small living organisms. Some examples are bacteria, fungi and yeast. They live in soil or water in every corner of the earth. Because they are so small, they are obviously a lot less complex than a calf. Even so, a simple microorganism can produce many different types of enzyme. In fact, it doesn't take a complex organism to create a complex enzyme-as was proved by a simple fungus named Mucor. This fungus produced a chymosin-like enzyme that was almost the same as that produced by calves. But finding the right enzyme is only half the battle.

Microorganisms often aren't Suited for "life on the Farm"

Once identified a microorganism needs to be "farmed" in a controlled setting. Most microorganisms can grow and multiply quite well in liquid. So large tanks are used called fermentation tanks to grow microorganisms. Ideally, a microorganism will grow fast and produce a lot of the desired enzyme at mild temperatures consuming inexpensive nutrients. But like most things in life, the ideal microorganism is hard to come by. As it is, most microorganisms found in the wild are not so well suited for domestication in fermentation tanks.

Some only produce tiny quantities of the desired enzyme, or they may produce undesirable by-products, or take a long time to grow, or require a growth environment that is very difficult and expensive to maintain. The result is that although we often find ourselves with a microorganism that can produce the enzyme needed the cost for cultivating it is prohibitive. That's where genetic engineering can help

Function and Structure

Enzymes are very efficient catalysts for biochemical reactions. They speed up reactions by providing an alternative reaction pathway of lower activation energy. Like all catalysts, enzymes take part in the reaction-that is how they provide an alternative reaction pathway. But they do not undergo permanent changes and so remain unchanged at the end of the reaction. They can only alter the rate of reaction, not the position of the equilibrium. Most chemical catalysts catalyse a wide range of reactions. They are not usually very selective. In contrast enzymes are usually highly selective, catalysing specific reactions only. This specificity is due to the shapes of the enzyme molecules. Many enzymes consist of a protein and a non-protein (called the cofactor). The proteins in enzymes are usually globular. The intra-and intermolecular bonds that hold proteins in their secondary and tertiary structures are disrupted by changes in temperature and pH. This affects shapes and so the catalytic activity of an enzyme is pH and temperature sensitive.

Cofactors may be:

- Organic groups that are permanently bound to the enzyme (prosthetic groups)
- Cations-positively charged metal ions (activators), which temporarily bind to the active site of the enzyme, giving an intense positive charge to the enzyme's protein
- Organic molecules, usually vitamins or made from vitamins (coenzymes), which are not permanently bound to the enzyme molecule, but combine with the enzyme-substrate complex temporarily.

HOW ENZYMES WORK

For two molecules to react they must collide with one another. They must collide in the right direction (orientation) and with sufficient energy. Sufficient energy means that between them they have enough energy to overcome the energy barrier to reaction. This is called the activation energy. Enzymes have an active site. This is part of the molecule that has just the right shape and functional groups to bind to one of the reacting molecules. The reacting molecule that binds to the enzyme is called the substrate. An enzyme-catalysed reaction takes a different 'route'. The enzyme and substrate form a reaction intermediate. Its formation has a lower activation energy than the reaction between reactants without a catalyst.

A simplified picture:

Route A reactant 1 + reactant 2 → product

Route B reactant 1 + enzyme → intermediate

intermediate + reactant 2 → product + enzyme

So the enzyme is used to form a reaction intermediate, but when this reacts with another reactant the enzyme reforms.

LOCK AND KEY HYPOTHESIS

This is the simplest model to represent how an enzyme works. The substrate simply fits into the active site to form a reaction intermediate.

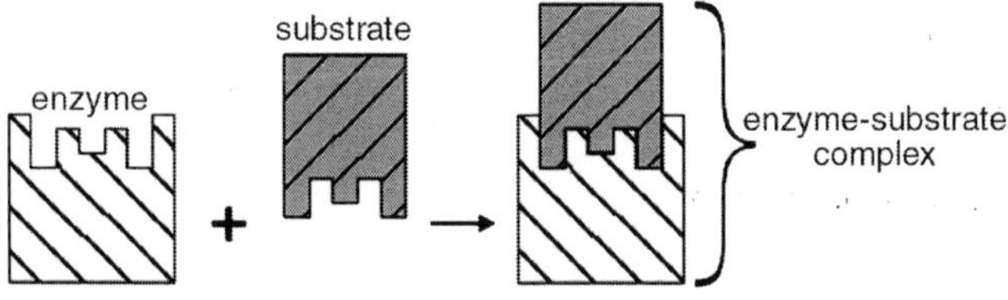

Induced Fit Hypothesis

In this model the enzyme molecule changes shape as the substrate molecules gets close. The change in shape is 'induced' by the approaching substrate molecule. This more sophisticated model relies on the fact that molecules are flexible because single covalent bonds are free to rotate.

Factors Affecting Catalytic Activity of Enzymes

Temperature

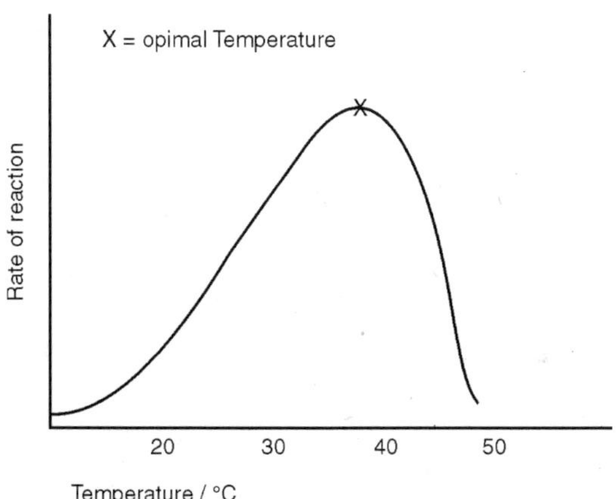

As the temperature rises, reacting molecules have more and more kinetic energy. This increases the chances of a successful collision and so the rate increases. There is a certain temperature at which an enzyme's catalytic activity is at its greatest. This optimal temperature is usually around human body temperature (37.5 °C) for the enzymes in human cells.

pH

Below this temperature the enzyme structure begins to break down (denature) since at higher temperatures intra-and intermolecular bonds are broken as the enzyme molecules gain even more kinetic energy. Each enzyme works within quite a small pH range. There is a pH at which its activity is greatest (the optimal pH). This is because changes in pH can make and break intra-and intermolecular bonds, changing the shape of the enzyme and, therefore, its effectiveness.

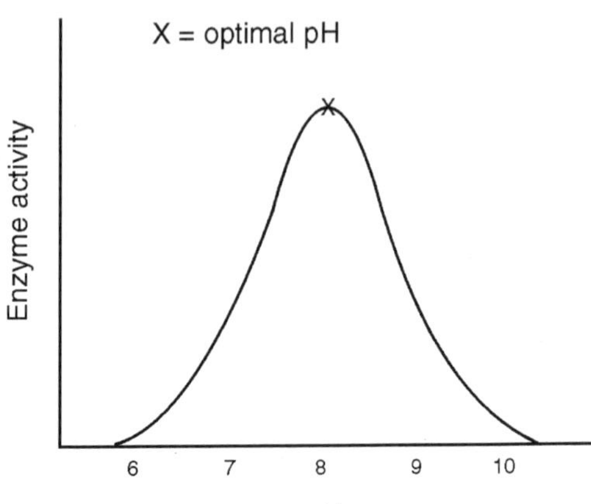

Concentration of Enzyme and Substrate

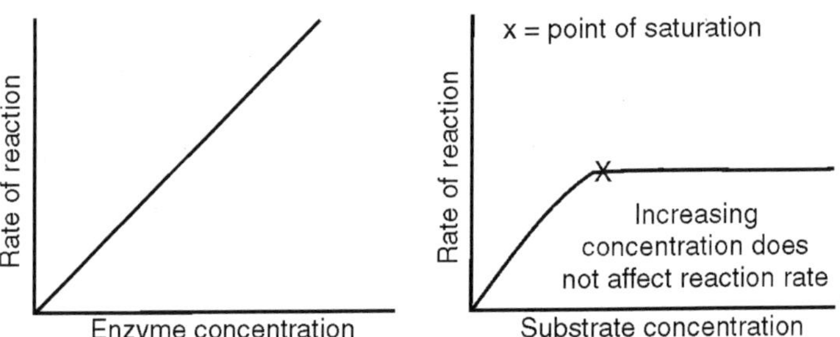

The rate of an enzyme-catalysed reaction depends on the concentrations of enzyme and substrate. As the concentration of either is increased the rate of reaction increases. For a given enzyme concentration, the rate of reaction increases with increasing substrate concentration up to a point, above which any further increase in substrate concentration produces no significant change in reaction rate.

This is because the active sites of the enzyme molecules at any given moment are virtually saturated with substrate. The enzyme/substrate complex has to dissociate before the active sites are free to accommodate more substrate Provided that the substrate concentration is high and that temperature and pH are kept constant, the rate of reaction is proportional to the enzyme concentration.

Inhibition of Enzyme Activity

Some substances reduce or even stop the catalytic activity of enzymes in biochemical reactions. They block or distort the active site. These chemicals are called inhibitors, because they inhibit reaction. Inhibitors that occupy the active site and prevent a substrate molecule from binding to the enzyme are said to be active site-directed (or competitive, as they 'compete' with the substrate for the active site). Inhibitors that attach to other parts of the enzyme molecule, perhaps distorting its shape, are said to be non-active site-directed (or non competitive).

Immobilized Enzymes

Enzymes are widely used commercially, for example in the detergent, food and brewing industries. Protease enzymes are used in 'biological' washing powders to speed up the breakdown of proteins in stains like blood and egg. Pectinase is used to produce and clarify fruit juices.

Problems using enzymes commercially include:

- They are water soluble which makes them hard to recover
- Some products can inhibit the enzyme activity (feedback inhibition) Enzymes can be immobilized by fixing them to a solid surface. This has a number of commercial advantages:
- The enzyme is easily removed
- The enzyme can be packed into columns and used over a long period
- Speedy separation of products reduces feedback inhibition
- Thermal stability is increased allowing higher temperatures to be used
- Higher operating temperatures increase rate of reaction

There are four principal methods of immobilization currently in use:

- Covalent bonding to a solid support
- Adsorption onto an insoluble substance
- Entrapment within a gel
- Encapsulation behind a selectively permeable membrane

2

Enzymes Action

INTRODUCTION

Enzymes are biological *catalysts* or assistants. Enzymes consist of various types of proteins that work to drive the chemical reaction required for a specific action or nutrient. Enzymes can either launch a reaction or speed it up. The chemicals that are transformed with the help of enzymes are called *substrates*. In the absence of enzymes, these chemicals are called *reactants*. To illustrate the speed and efficiency of enzymes, substrates can be transformed to usable products at the rate of ten times per second. Considering that there are an estimated 75,000 different enzymes in the human body, these chemical reactions are performed at an amasing rate. On the other hand, in the absence of enzymes, reactants may take hundreds of years to convert into a usable product, if they are able to do so at all. This is why enzymes are crucial in the sustenance of life on earth. Generally, enzymes work on substrates in one of three ways: substrate orientation, physical stress, and changes in substrate reactivity. Substrate orientation occurs when an enzyme causes substrate molecules to align with each other and form a bond. When an enzyme uses physical stress on a substrate, it in effect grips the substrate and forces the molecule to break apart. An enzyme that causes changes in substrate reactivity alters the placement of the molecule's electrons, which influences the molecule's ability to bond with other molecules. Enzymes have active sites where they come into contact with particular substrates. The catalytic properties of enzymes are a cyclic process. Once a substrate has come into contact with the active site of an enzyme, it is modified by the enzyme to form the end product.

Once the process is complete, the enzyme releases the product and is ready to begin the process with new substrates. Enzymes are never wasted and always recycled. The absence of enzymes is responsible for many diseases. In humans, a tragic disease called phenylketonuria (PKU), which causes severe mental retardation and even death in infants, is the result of the absence of one type of enzyme. Tay-Sachs disease is a similarly tragic result of an enzyme deficiency. It causes retardation, paralysis, and often death in early childhood when left untreated.

REGULATION OF ENZYME ACTIVITY

Several mechanisms work to make enzyme activity within the cell efficient and well-coordinated.

Anchoring enzymes in membranes

Many enzymes are inserted into cell membranes, for examples,

- The plasma membrane
- The membranes of mitochondria and chloroplasts
- The endoplasmic reticulum
- The nuclear envelope

These are locked into spatial relationships that enable them to interact efficiently.

Inactive Precursors

Enzymes, such as proteases, that can attack the cell itself are inhibited while within the cell that synthesises them. For example, pepsin is synthesised within thechief cells (in gastric glands) as an inactive precursor, pepsinogen. Only when exposed to the low pH outside the cell is the inhibiting portion of the molecule removed and active pepsin produced.

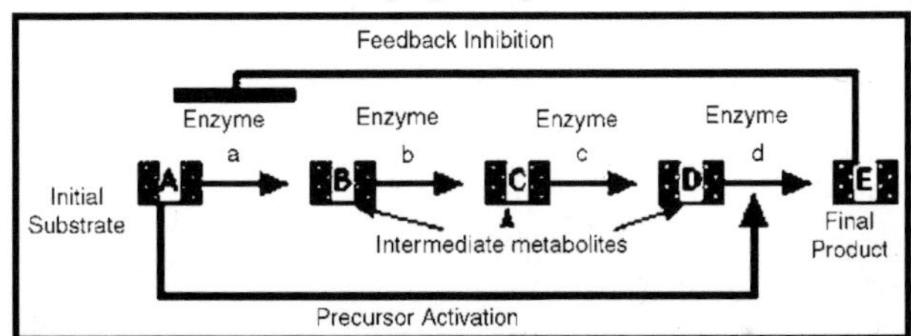

Feedback Inhibition

If the product of a series of enzymatic reactions, *e.g.*, an amino acid, begins to accumulate within the cell, it may specifically inhibit the action of the first enzyme involved in its synthesis (red bar). Thus further production of the enzyme is halted.

Precursor Activation

The accumulation of a substance within a cell may specifically activate (blue arrow) an enzyme that sets in motion a sequence of reactions for which that substance is the initial substrate. This reduces the concentration of the initial substrate.

In the case if feedback inhibition and precursor activation, the activity of the enzyme is being regulated by a molecule which is not its substrate. In these cases, the regulator molecule binds to the enzyme at a different site than the one to which the substrate binds. When the regulator binds to its site, it alters the shape of the enzyme so that its activity is changed. This is called an allosteric effect.

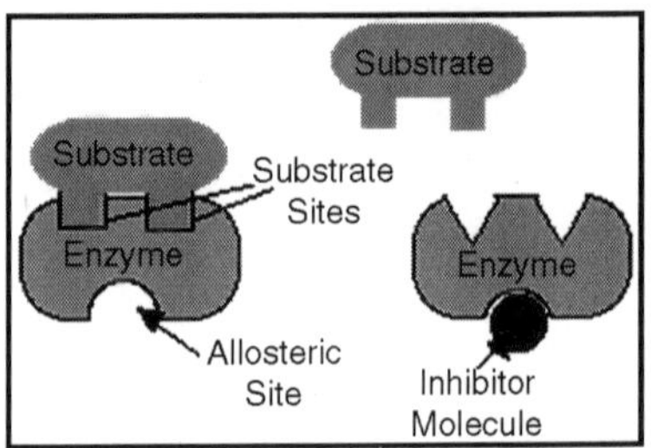

- In feedback inhibition, the allosteric effect lowers the affinity of the enzyme for its substrate.
- In precursor activation, the regulator molecule increases the affinity of the enzyme in the series for its substrate.

REGULATION OF ENZYME SYNTHESIS

The four mechanisms described above regulate the activity of enzymes already present within the cell. What about enzymes that are not needed or are needed but not present? Here, too, control mechanisms are at work that regulate the rate at which new enzymes are synthesised. Most of these controls work by turning on — or off — the transcription of genes. If, for example, ample quantities of an amino acid are already available to the cell from its extracellular fluid, synthesis of the enzymes that would enable the cell to produce that amino acid for itself is shut down.

Conversely, if a new substrate is made available to the cell, it may induce the synthesis of the enzymes needed to cope with it. Yeast cells, for example, do not ordinarily metabolise lactose and no lactase can be detected in them. However, if grown in a medium containing lactose, they soon begin synthesising lactase — by transcribing and translating the necessary gene(s) — and so can begin to metabolise the sugar.

BIOLOGICAL CATALYSTS OF ENZYMES

This means that they speed up the chemical reactions in living things. Without enzymes, our guts would take weeks and weeks to digest our food, our muscles, nerves and bones would not work properly and so on - we would not be living!

A catalyst is any substance which makes a chemical reaction go faster, without itself being changed. A catalyst can be used over and over again in a chemical reaction: it does not get used up. Enzymes are very much the same except that they are easily denatured (destroyed: but do NOT use this word since the protein molecule is not broken down into amino-acids, it just loses it shape and will not work any more) by heat. Our enzymes work best at body temperature. Our enzymes also have to have the correct pH. pH and heavy metal ions. Unlike ordinary catalysts, they are specific to one chemical reaction. An ordinary catalyst may be used for several different chemical reactions, but an enzyme only works for one specific reaction.

Human saliva contains an enzyme called amylase. This enzyme helps to turn starch into a sugar called maltose. When you swallow a mouthful of food, the amylase stops working because it is much too acid in the stomach pH 2. Amyalse works best in neutral or slightly alkaline conditions, *i.e.*, at about pH 7. When your food gets into the small intestine, more amylase is made by the pancreas and this turns the remaining starch into maltose. Another enzyme (maltase) turns all this maltose into glucose. Glucose is then absorbed into the blood. Enzymes in the human alimentary canal and what they digest:

Enzyme	Substrate
Amylase	Starch
Maltase	Maltose
Sucrase	Sucrose
Lipase	Fats
Pepsin	Proteins

All animals, green plants, fungi and bacteria produce enzymes: so enzymes are not just about digesting food. The enzymes which we use to digest our food are extra-cellular, that means they are found outside cells. We also have enzymes inside our cells; these are intra-cellular enzymes. Enzymes are used in ALL chemical reactions in living things; this includes respiration, photosynthesis, movement growth, getting rid of toxic chemicals in the liver and so on. Viruses are rather different, but you do not need to know much about them for GCSE, so just make sure that you don't catch any! Enzymes must have the correct shape to do their job. They are made of proteins, and proteins are very easily affected by heat, pH and heavy metal ions. Some people say that enzymes work like a key in a lock.

If the key has been twisted by heat, or dissolved in acid or stuck up with chewing gum it will not work. Enzymes change their shape if the temperature or pH changes, so they have to have the right conditions. Copper ions are poisonous: if you get copper ions in your blood they will block up some of the important enzymes in red and white blood cells. Enzymes are catalysts. Most are proteins. (A few ribonucleo-protein enzymes have been discovered and, for some of these, the catalytic activity is in the RNA part rather than the protein part. Link to discussion of theseribozymes.) Enzymes bind temporarily to one or more of the reactants of the reaction they catalyse. In doing so, they lower the amount of activation energy needed and thus speed up the reaction.

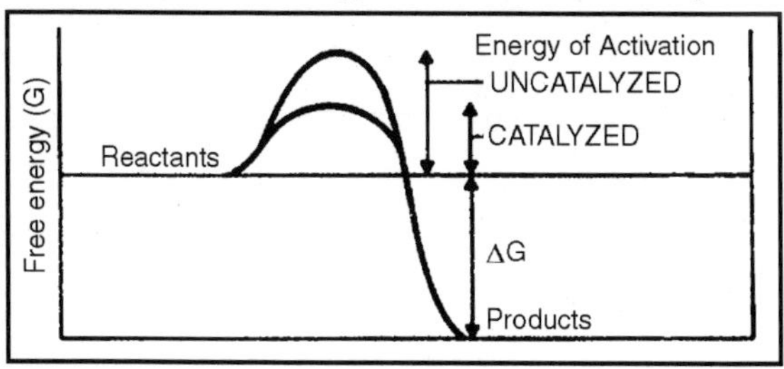

Examples:

- *Catalase*: It catalyses the decomposition of hydrogen peroxide into water and oxygen.

$$2H_2O_2 \rightarrow 2H_2O + O_2$$

One molecule of catalase can break 40 million molecules of hydrogen peroxide each second.

- *Carbonic anhydrase*: It is found in red blood cells where it catalyses the reaction

$$CO_2 + H_2O \leftrightarrow H_2CO_3$$

It enables red blood cells to transport carbon dioxide from the tissues to the lungs. One molecule of carbonic anhydrase can process one million molecules of CO_2 each second.

- *Acetylcholinesterase:* It catalyses the breakdown of the neurotransmitter acetylcholine at several types of synapses as well as at the neuromuscular junction — the specialised synapse that triggers the contraction of skeletal muscle.

One molecule of acetylcholinesterase breaks down 25,000 molecules of acetylcholine each second. This speed makes possible the rapid "resetting" of the synapse for transmission of another nerve impulse. In order to do its work, an enzyme must unite — even if ever so briefly — with at least one of the reactants. In most cases, the forces that hold the enzyme and its substrate are non-covalent, an assortment of:

- hydrogen bonds
- ionic interactions
- and hydrophobic interactions

Most of these interactions are weak and especially so if the atoms involved are farther than about one angstromfrom each other. So successful binding of enzyme and substrate requires that the two molecules be able to approach each other closely over a fairly broad surface. Thus the analogy that a substrate molecule binds its enzyme like a key in a lock. This requirement for complementarity in the configuration of substrate and enzyme explains the rema-rkable specificity of most enzymes. Generally, a given enzyme is able to catalyse only a single chemical reaction or, at most, a few reactions involving substrates sharing the same general structure.

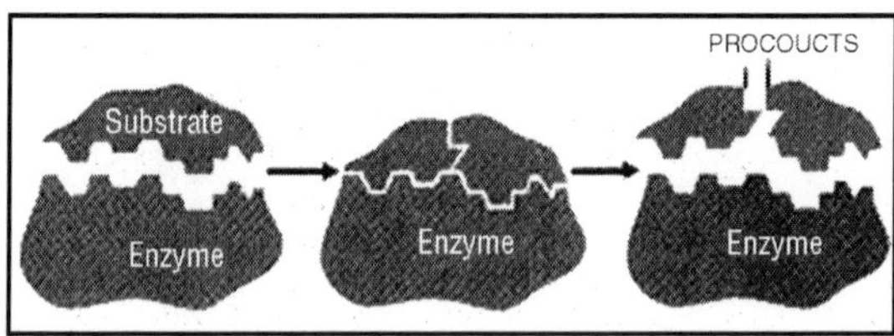

COMPETITIVE INHIBITION

The necessity for a close, if brief, fit between enzyme and substrate explains the phenomenon of competitive inhibition.

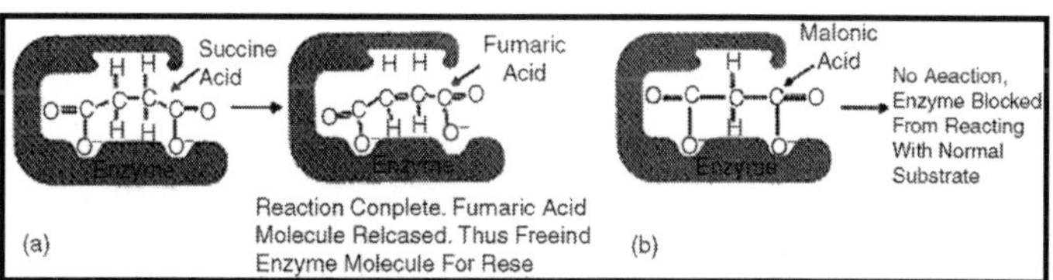

One of the enzymes needed for the release of energy within the cell is succinic dehydrogenase. It catalyses the oxidation (by the removal of two hydrogen atoms) of succinic acid (a). If one adds malonic acid to cells, or to a test tube mixture of succinic acid and the enzyme, the action of the enzyme is strongly inhibited. This is because the structure of malonic acid allows it to bind to the same site on the enzyme(b). But there is no oxidation so no speedy release of products. The inhibition is called competitive because if you increase the ratio of succinic to malonic acid in the mixture, you will gradually restore the rate of catalysis. At a 50:1 ratio, the two molecules compete on roughly equal terms for the binding (=catalytic) site on the enzyme.

CELLS ORGANISMS OF ENZYMES AS BIOLOGICAL CATALYSTS

In cells and organisms most reactions are catalyzed by enzymes, which are regenerated during the course of a reaction. These biological catalysts are physiologically important because they speed up the rates of reactions that would otherwise be too slow to support life. Enzymes increase reaction rates, sometimes by as much as one million-fold, but more typically by about one thousand fold. Catalysts speed up the forward and reverse reactions proportionately so that, although the magnitude of the rate constants of the forward and reverse reactions is are increased, the ratio of the rate constants remains the same in the presence or absence of enzyme. Since the equilibrium constant is equal to a ratio of rate constants, it is apparent that enzymes and other catalysts have no effect on the equilibrium constant of the reactions they catalyze.

Enzymes increase reaction rates by decreasing the amount of energy required to form a complex of reactants that is competent to produce reaction products. This complex is known as the activated state or transition state complex for the reaction. Enzymes and other catalysts accelerate reactions by lowering the energy of the transition state. The free energy required to form an activated complex is much lower in the catalyzed reaction. The amount of energy required to achieve the transition state is lowered; consequently, at any instant a greater proportion of the molecules in the population can achieve the transition state. The result is that the reaction rate is increased.

Michaelis-Menten Kinetics

In typical enzyme-catalyzed reactions, reactant and product concentrations are usually hundreds or thousands of times greater than the enzyme concentration. Consequently, each enzyme molecule catalyzes the conversion to product of many reactant molecules. In biochemical reactions, reactants are commonly known as substrates. The catalytic event that converts substrate to product involves the formation of a transition state, and it occurs most easily at a specific binding site on the enzyme. This site, called the catalytic site of the enzyme, has been evolutionarily structured to provide specific, high-affinity binding of substrate(s) and to provide an environment that favours the catalytic events. The complex that forms, when substrate(s) and enzyme combine, is called the enzyme substrate (ES) complex. Reaction products arise when the ES complex breaks down releasing free enzyme.

Between the binding of substrate to enzyme, and the reappearance of free enzyme and product, a series of complex events must take place. At a minimum an ES complex must be formed; this complex must pass to the transition state (ES*); and the transition state complex must advance to an enzyme product complex (EP). The latter is finally competent to dissociate to product and free enzyme.

The series of events can be shown thus:

$$E + S \leftrightarrow ES \leftrightarrow ES^* \leftrightarrow EP \leftrightarrow E + P$$

The kinetics of simple reactions like that above were first characterized by biochemists Michaelis and Menten. The concepts underlying their analysis of enzyme kinetics continue to provide the cornerstone for understanding metabolism today, and for the development and clinical use of drugs aimed at selectively altering rate constants and interfering with the progress of disease states. The Michaelis-Menten equation is a quantitative description of the relationship among the rate of an enzyme- catalyzed reaction [v_1], the concentration of substrate [S] and two constants, V_{max} and K_m (which are set by the particular equation). The symbols used in the Michaelis-Menten equation refer to the reaction rate [v_1], maximum reaction rate (V_{max}), substrate concentration [S] and the Michaelis-Menten constant (K_m).

$$V_1 = \frac{V_{max}[s]}{\{K_m + [S]\}}$$

The Michaelis-Menten equation can be used to demonstrate that at the substrate concentration that produces exactly half of the maximum reaction rate, *i.e.* ½ V_{max}, the substrate concentration is numerically equal to K_m. This fact provides a simple yet powerful bioanalytical tool that has been used to characterize both normal and altered enzymes, such as those that produce the symptoms of genetic diseases. Rearranging the Michaelis-Menten equation leads to:

$$K_m = [S]\left\{\left[\frac{V_{max}}{V_1}\right] - 1\right\}$$

From this equation it should be apparent that when the substrate concentration is half that required to support the maximum rate of reaction, the observed rate, v_1, will, be equal to V_{max} divided by 2; in other words, $v_1 = [V_{max}/2]$. At this substrate concentration V_{max}/v_1 will be exactly equal to 2, with the result that:

$$[S](1) = K_m$$

The latter is an algebraic statement of the fact that, for enzymes of the Michaelis-Menten type, when the observed reaction rate is half of the maximum possible reaction rate, the substrate concentration is numerically equal to the Michaelis-Menten constant. In this derivation, the units of K_m are those used to specify the concentration of S, usually Molarity. The Michaelis-Menten equation has the same form as the equation for a rectangular hyperbola; graphical analysis of reaction rate (v) versus substrate concentration [S] produces a hyperbolic rate plot.

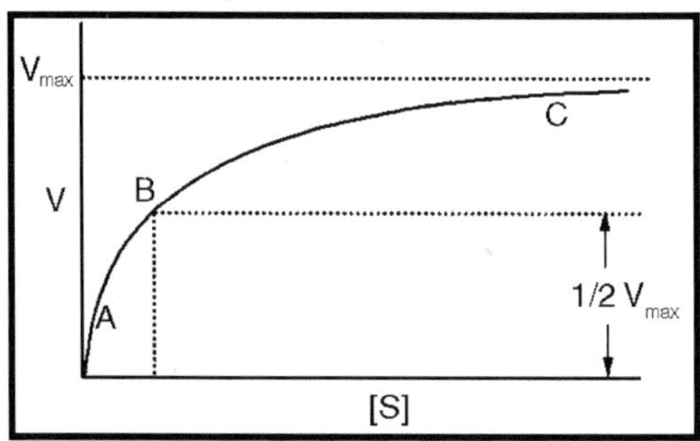

Fig. Plot of Substrate Concentration Versus Reaction Velocity

The key features of the plot are marked by points A, B and C. At high substrate concentrations the rate represented by point C the rate of the reaction is almost equal to V_{max}, and the difference in rate at nearby concentrations of substrate is almost negligible.

If the Michaelis-Menten plot is extrapolated to infinitely high substrate concentrations, the extrapolated rate is equal to V_{max}. When the reaction rate becomes independent of substrate concentration, or nearly so, the rate is said to be zero order. (Note that the reaction is zero order only with respect to this substrate. If the reaction has two substrates, it may or may not be zero order with respect to the second substrate).

The very small differences in reaction velocity at substrate concentrations around point C (near V_{max}) reflect the fact that at these concentrations almost all of the enzyme molecules are bound to substrate and the rate is virtually independent of substrate, hence zero order. At lower substrate concentrations, such as at points A and B, the lower reaction velocities indicate that at any moment only a portion of the enzyme molecules are bound to the substrate. In fact, at the substrate concentration denoted by point B, exactly half the enzyme molecules are in an ES complex at any instant and the rate is exactly one half of V_{max}. At substrate concentrations near point A the rate appears to be directly proportional to substrate concentration, and the reaction rate is said to be first order.

Inhibition of Enzyme Catalyzed Reactions

To avoid dealing with curvilinear plots of enzyme catalyzed reactions, biochemists Lineweaver and Burk introduced an analysis of enzyme kinetics based on the following rearrangement of the Michaelis-Menten equation:

$$\frac{1}{V} = \left[\frac{K_m(1)}{V_{max}(S)} + \frac{1}{V_{max}}\right]$$

Plots of 1/v versus 1/[S] yield straight lines having a slope of K_m/V_{max} and an intercept on the ordinate at $1/V_{max}$.

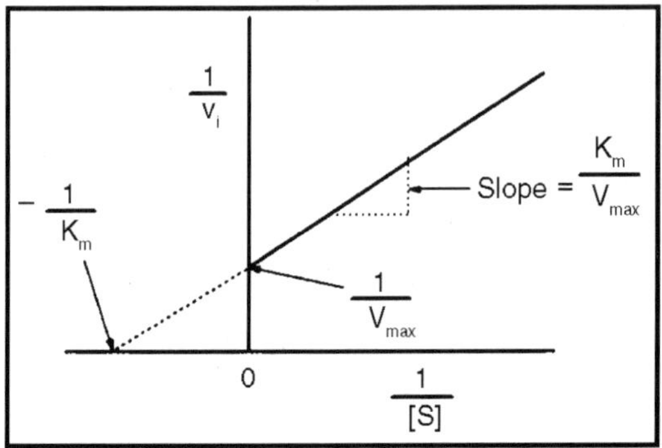

Fig. A Lineweaver-Burk Plot

An alternative linear transformation of the Michaelis-Menten equation is the Eadie-Hofstee transformation:

$$\frac{v}{[S]} = -v\left[\frac{1}{K_m} +\right] + \left[\frac{V_{max}}{K_m}\right]$$

and when v/[S] is plotted on the y-axis versus v on the x-axis, the result is a linear plot with a slope of $-1/K_m$ and the value V_{max}/K_m as the intercept on the y-axis and V_{max} as the intercept on the x-axis.

Both the Lineweaver-Burk and Eadie-Hofstee transformation of the Michaelis-Menton equation are useful in the analysis of enzyme inhibition. Well- known examples of such therapy include the use of methotrexate in cancer chemotherapy to semi-selectively inhibit DNA synthesis of malignant cells, the use of aspirin to inhibit the synthesis of prostaglandins which are at least partly responsible for the aches and pains of arthritis, and the use of sulfa drugs to inhibit the folic acid synthesis that is essential for the metabolism and growth of disease-causing bacteria. In addition, many poisons, such as cyanide, carbon monoxide and polychlorinated biphenols (PCBs) produce their life-threatening effects by means of enzyme inhibition. Enzyme inhibitors fall into two broad classes: those causing irreversible inactivation of enzymes and those whose inhibitory effects can be reversed. Inhibitors of the first class usually cause an inactivating, covalent modification of enzyme structure. Cyanide is a classic example of an irreversible enzyme inhibitor: by covalently binding mitochondrial cytochrome oxidase, it inhibits all the reactions associated with electron transport.

The kinetic effect of irreversible inhibitors is to decrease the concentration of active enzyme, thus decreasing the maximum possible concentration of ES complex. Since the limiting enzyme reaction rate is often k_2[ES], it is clear that under these circumstances the reduction of enzyme concentration will lead to decreased reaction rates.

Note that when enzymes in cells are only partially inhibited by irreversible inhibitors, the remaining unmodified enzyme molecules are not distinguishable from those in untreated cells; in particular, they have the same turnover number and the same K_m. Turnover number, related to V_{max}, is defined as the maximum number of moles of substrate that can be converted to product per mole of catalytic site per second. Irreversible inhibitors are usually considered to be poisons and are generally unsuitable for therapeutic purposes. Reversible inhibitors can be divided into two main categories; competitive inhibitors and noncompetitive inhibitors, with a third category, uncompetitive inhibitors, rarely encountered.

Inhibitor Type	Binding Site on Enzyme	Kinetic effect
Competitive Inhibitor	Specifically at the catalytic site, where it competes with substrate for binding in a dynamic equilibrium-like process. Inhibition is reversible by substrate.	V_{max} is unchanged; K_m, as defined by [S] required for ½ maximal activity, is increased.
Noncompetitive Inhibitor	Binds E or ES complex other than at the catalytic site. Substrate binding unaltered, but ESI complex cannot form products. Inhibition cannot be reversed by substrate. concentration.	K_m appears unaltered; V_{max} is decreased proportionately to inhibitor
Uncompetitive Inhibitor	Binds only to ES complexes at locations other than the catalytic site. Substrate binding modifies enzyme structure, making inhibitor- binding site available. Inhibition cannot be reversed by substrate.	Apparent V_{max} decreased; K_m, as defined by [S] required for ½ maximal activity, is decreased.

The hallmark of all the reversible inhibitors is that when the inhibitor concentration drops, enzyme activity is regenerated. Usually these inhibitors bind to enzymes by non-covalent forces and the inhibitor maintains a reversible equilibrium with the enzyme. The equilibrium constant for the dissociation of enzyme inhibitor complexes is known as K_i:

$$K_i = \frac{[E][I]}{[EI]}$$

The importance of K_I is that in all enzyme reactions where substrate, inhibitor and enzyme interact, the normal K_m and or V_{max} for substrate enzyme interaction appear to be altered. These changes are a consequence of the influence of K_i on the overall rate equation for the reaction. The effects of K_i are best observed in Lineweaver-Burk plots. Probably the best known reversible inhibitors are competitive inhibitors, which always bind at the catalytic or active site of the enzyme. Most drugs that alter enzyme activity are of this type.

Competitive inhibitors are especially attractive as clinical modulators of enzyme activity because they offer two routes for the reversal of enzyme inhibition, while other reversible inhibitors offer only one. First, as with all kinds of reversible inhibitors, a decreasing concentration of the inhibitor reverses the equilibrium regenerating active free enzyme. Second, since substrate and competitive inhibitors both bind at the same site they compete with one another for binding

Raising the concentration of substrate (S), while holding the concentration of inhibitor constant, provides the second route for reversal of competitive inhibition. The greater the proportion of substrate, the greater the proportion of enzyme present in competent ES complexes. As noted earlier, high concentrations of substrate can displace virtually all competitive inhibitor bound to active sites. Thus, it is apparent that V_{max} should be unchanged by competitive inhibitors.

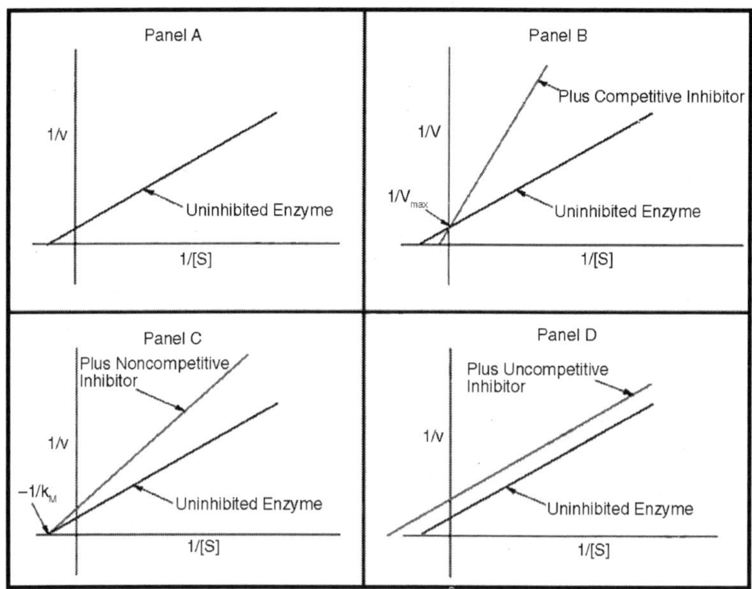

Fig. Lineweaver-Burk Plots of Inhibited Enzymes

This characteristic of competitive inhibitors is reflected in the identical vertical-axis intercepts of Lineweaver-Burk plots, with and without inhibitor. Since attaining V_{max} requires appreciably higher substrate concentrations in the presence of competitive inhibitor, K_m (the substrate concentration at half maximal velocity) is also higher, as demonstrated by the differing negative intercepts on the horizontal axis in panel B.

Analogously, panel C illustrates that noncompetitive inhibitors appear to have no effect on the intercept at the x-axis implying that noncompetitive inhibitors have no effect on the K_m of the enzymes they inhibit. Since noncompetitive inhibitors do not interfere in the equilibration of enzyme, substrate and ES complexes, the K_m's of Michaelis-Menten type enzymes are not expected to be affected by noncompetitive inhibitors, as demonstrated by x-axis intercepts in panel C. However, because complexes that contain inhibitor (ESI) are incapable of progressing to reaction products, the effect of a noncompetitive inhibitor is to reduce the concentration of ES complexes that can advance to product. Since $V_{max} = k_2[E_{total}]$, and the concentration of competent E_{total} is diminished by the amount of ESI formed, noncompetitive inhibitors are expected to decrease V_{max}, as illustrated by the y-axis intercepts in panel C.

A corresponding analysis of uncompetitive inhibition leads to the expectation that these inhibitors should change the apparent values of K_m as well as V_{max}. Changing both constants leads to double reciprocal plots, in which intercepts on the x and y axes are proportionately changed; this leads to the production of parallel lines in inhibited and uninhibited reactions.

ENZYME COFACTORS

Many enzymes require the presence of an additional, non-protein, cofactor.

- Some of these are metal ions such as Zn^{2+} (the cofactor for carbonic anhydrase), Cu^{2+}, Mn^{2+}, K^+, and Na^+.
- Some cofactors are small organic molecules called coenzymes. The B vitamins
  - thiamine (B1)
  - riboflavin (B2) and
  - nicotinamide are precursors of coenzymes.

Coenzymes may be covalently bound to the protein part (called the apoenzyme) of enzymes as a prosthetic group. Others bind more loosely and, in fact, may bind only transiently to the enzyme as it performs its catalytic act.

LYSOZYME: A MODEL OF ENZYME ACTION

A number of lysozymes are found in nature; in human tears and egg white, for examples. The enzyme is antibacterial because it degrades the polysaccharide that is found in the cell walls of many bacteria. It does this by catalysing the insertion of a water molecule at the position indicated by the red arrow (aglycosidic bond). This hydrolysis breaks the chain at that point.

The bacterial polysaccharide consists of long chains of alternating amino sugars:

- N-acetylglucosamine (NAG)
- N-acetylmuramic acid (NAM)

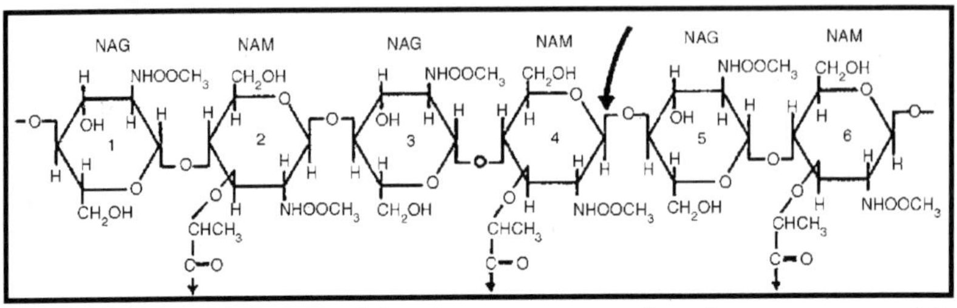

These hexose units resemble glucose except for the presence of the side chains containing amino groups.

Lysozyme is a globular protein with a deep cleft across part of its surface. Six hexoses of the substrate fit into this cleft.

- With so many oxygen atoms in sugars, as many as 14 hydrogen bonds form between the six amino sugars and certain amino acid R groups such as Arg-114, Asn-37, Asn-44, Trp-62, Trp-63, and Asp-101.
- Some hydrogen bonds also form with the C=O groups of several peptide bonds.
- In addition, hydrophobic interactions may help hold the substrate in position.

X-ray crystallography has shown that as lysozyme and its substrate unite, each is slightly deformed. The fourth hexose in the chain (ring #4) becomes twisted out of its normal position.

This imposes a strain on the C-O bond on the ring-4 side of the oxygen bridge between rings 4 and 5. It is just at this point that the polysaccharide is broken. A molecule of water is inserted between these two hexoses, which breaks the chain. Here, then, is a structural view of what it means to lower activation energy. The energy needed to break this covalent bond is lower now that the atoms connected by the bond have been distorted from their normal position.

As for lysozyme itself, binding of the substrate induces a small (~0.75Å) movement of certain amino acid residues so the cleft closes slightly over its substrate. So the "lock" as well as the "key" changes shape as the two are brought together. (This is sometimes called "induced fit".)

The amino acid residues in the vicinity of rings 4 and 5 provide a plausible mechanism for completing the catalytic act. Residue 35, glutamic acid (Glu-35), is about 3Å from the -O- bridge that is to be broken. The free carboxyl group of glutamic acid is a hydrogen ion donor and available to transfer H^+ to the oxygen atom. This would break the already-strained bond between the oxygen atom and the carbon atom of ring 4. Now having lost an electron, the carbon atom acquires a positive charge. Ionised carbon is normally very unstable, but the attraction of the negatively-charged carboxyl ion of Asp-52 could stabilise it long enough for an -OH ion (from a spontaneously dissociated water molecule) to unite with the carbon.

Even at pH 7, water spontaneously dissociates to produce H^+ and OH^- ions.] The hydrogen ion (H^+) left over can replace that lost by Glu-35.

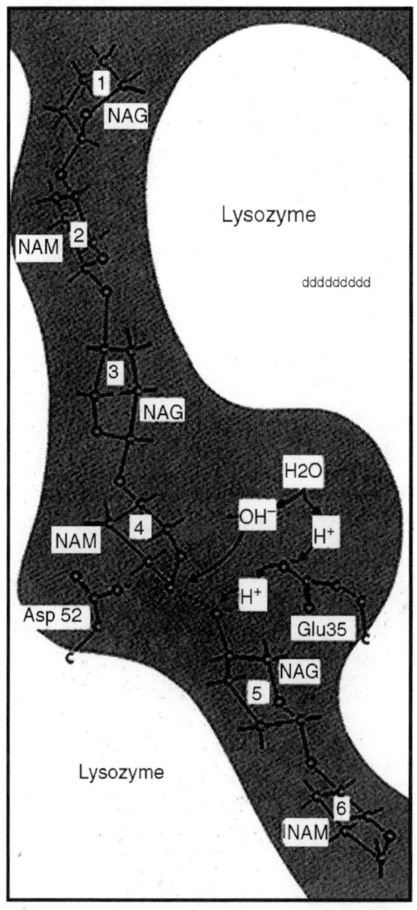

In either case, the chain is broken, the two fragments separate from the enzyme, and the enzyme is free to attach to a new location on the bacterial cell wall and continue its work of digesting it.

FACTORS AFFECTING ENZYME ACTION

The activity of enzymes is strongly affected by changes in pH and temperature. Each enzyme works best at a certain pH (left graph) and temperature (right graph), its activity decreasing at values above and below that point. This is not surprising considering the importance of

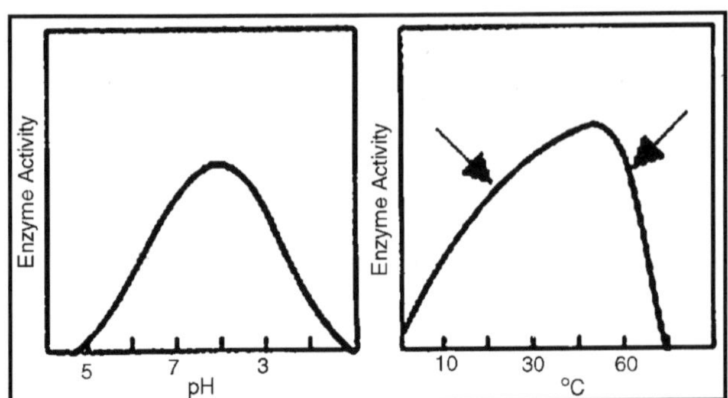

- Tertiary structure (*i.e.*, shape) in enzyme function and
- Non-covalent forces, *e.g.*, ionic interactions and hydrogen bonds, in determining that shape.

Examples:

- The protease pepsin works best as a pH of 1–2 (found in the stomach) while
- The protease trypsin is inactive at such a low pH but very active at a pH of 8 (found in the small intestine as the bicarbonate of the pancreatic fluid neutralises the arriving stomach contents).

Changes in pH alter the state of ionisation of charged amino acids (*e.g.*, Asp, Lys) that may play a crucial role in substrate binding and/or the catalytic action itself. Without the unionised -COOH group of Glu-35 and the ionised -COO⁻ of Asp-52, the catalytic action of lysozyme would cease.

Hydrogen bonds are easily disrupted by increasing temperature. This, in turn, may disrupt the shape of the enzyme so that its affinity for its substrate diminishes. The ascending portion of the temperature curve (red arrow in right-hand graph above) reflects the general effect of increasing temperature on the rate of chemical reactions (graph at left). The descending portion of the curve above reflects the loss of catalytic activity as the enzyme molecules become denatured at high temperatures.

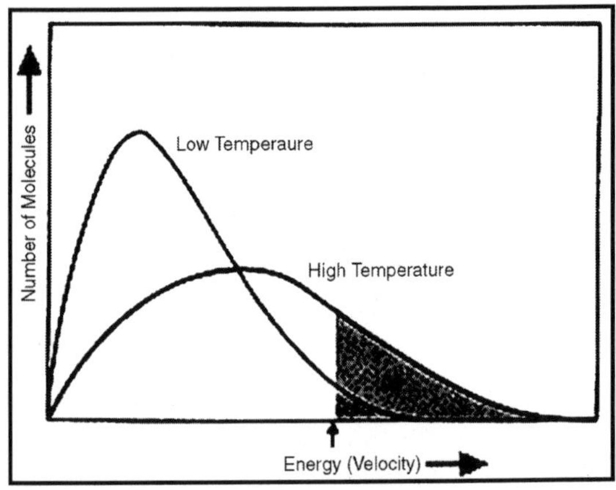

THE GENERAL PROPERTIES OF ENZYMES

All enz)niies are colloids, like the native proteins of the seralbumin type. Like these, they cannot be dialysed through parchment, and are irrevocably changed on heating in aqueous solution to 60° to 70° C. The albumin is coagulated by heat, and this coagulum cannot be dissolved without conversion into peptones; the enzymes, however, disappear entirely on heating. It is possible that they are coagulated and become insoluble; it is also possible that they are chemically changed and decomposed. The only sign of the presence of enzymes is their action, and we see that the hydrolysing or oxidising power of a solution is lost by heating, and cannot be restored. Pfeflfer was able to show that the enzymes do not volatiUze with the vapor of water and thus do not leave the solution.

All enzymes are destroyed by heat. Most of them are more or less slowly impaired at a low temperature 37° to 40° C. – ^and many undergo a slow loss of power, at room temperature, and even at zero. Zymase and the glycolytic enzyme of muscles are destroyed in the frozen state in two to four days.^ Enzymes are also broken down by many chemical processes, by strong acids and alkalies, by alcohol, etc. Pepsin loses its enzymotic properties by contact with the weakest alkali, and many oxidising enzymes by acid. It is possible that this alteration is the same as that which takes place at a slower rate on heating. Bayliss* supposes that the alteration at low temperature is reversible. He cites as an example, that toxins may become inactive toxoids while retaining their chemical character.

This supposition was re- cently supported in a very remarkable manner by Pawlow.' It has always been believed that the pepsin is completely destroyed by alkali; Pawlow and Tichomirow demonstrat- ed that this is the case only if the alkaline solution is im mediately acidified; the pepsin cannot then be detected. If, however, the gastric juice be made alkaline, then neutralised, and allowed to stand for some time before acidulating it, the pepsin does not disappear, but acts almost as strongly as before. This is the first case of restitution of an enzyme after having lost its vigour. Melt- zer has pointed out that this slow loss of power is greatly accelerated on shaking solutions of enzymes. The de- struction through shaking takes place even at low tem- peratures, but proceeds more rapidly at the temperature of the body.

To-day every investigation of enzymes is retarded by the same difficulties as those attending the chemical study of proteins. The true proteins are like enzymes so far as sensitiveness towards temperature and towards acid and alkali are concerned. Knowledge of the chemistry of the proteins stagnated for a long time, because the methods of dealing with colloids were undeveloped. It made very rapid progress as soon as it was found possible to work with strong hydrochloric acid, barium hydroxide, and phosphotungstic acid. I have cited the methods of purifying enzymes, salting out, and precipitating by weak acid or alcohol, avoiding heat. By such methods we are unable to separate compounds which can be chemically defined. It is to be hoped that Pawlow's observation may become the beginning of a really chemical explora- tion of enzymes.

' We cannot detect or estimate the enzymes themselves. We can only observe their action. For example, we can estimate the quantity of cane sugar before and after the ac- tion of invertin; we can also determine the quantity of glucose and levulose, the products of the decomposition of the cane sugar. Further, we can estimate the quantity of the coagulable proteins before the action, and after five, ten, and fifteen minutes, or the quantity of the non-coagulable peptone; but we cannot directly estimate the invertin or pepsin.

Therefore everything that disturbs the action of the enzyme seems to be a diminution, while everytlnng that improves the action seems to be an increase, of the quantity of the enzyme, although possibly only due to differences in the conditions of its action. rule, generally acknowledged in chemistry, that for deternjining the physical properties of a substance - ^the molecular weight, the melting-point, the osmotic pressure, the rotatory power, the electrical conductivity, the ioniza- tion - ^the substance must be wholly purified. We know that in enyzme solutions the weight of enzymes is much smaller than the weight of the impurities; it is clear that in such a solution we cannot define the law of action of these bodies. If we read the books or papers written on enzymes by chemists, we find that the law of Schultz states that the action of a solution of an enzyme increases, not in proportion to the quantity, but to the square root of the quantity of the contained enzyme; and the question is also, debated, whether the enzyme is consumed during its action or not, etc. The prevailing great interest in the enzymes has led authors to bring up questions which no body can answer at the present time. All we can do at present is to describe a number of the properties of enzymes which have been observed.

- Enzymes are colloids, and do not dialyse through parchment.
- Enzymes have an optimum temperature of action. It is well known that all chemical processes go faster with increasing temperature, and van 't Hoff enunciated the rule that the rapidity of most chemical processes is doubled or trebled when the temperature is raised io° C. The enzymes follow this rule, but only between o° and 40° C. above 40° C, this power decreases rapidly. If we graphi cally represent the behaviour of the decomposition of canesugar by hydrochloric acid and by invertin, with increasing temperature, we find for the acid a curve like this^^^^, and for the enzyme a curve/ ^, *i.e.*, a curve with a dis- tinct maximum. It is possible that two processes are involved: the general increasing reaction with rise of temperature, and the destruction of the enzymes by heat. But this is not certain. In some enzymes we can see a slow acceleration of action at from io° to 30° C, and a much more rapid acceleration at from

30° to 37° C. The curve has a form difficult to explain as the summation of opposed effects. It is possible that we have, besides the general acceleration with rising temperature, a special adaptation of the enzymes to the body-temperature of the higher animals. The optimum of most of the enzymes examined lies between 30° and 40° C. But we find old statements that the optimum of the diastase of yeast lies at 50° C, or that the pepsin of fish has an optimum between 2° and 5° C. These statements do not hold good, as the tissues of cold-blooded animals contain proteins coagulat- ing spontaneously at 30° to 40° C; if these proteins are precipitated, pepsin is precipitated by adsorption, and the extract becomes accordingly poorer without change of the pepsin itself. In the case of yeast, the better solubility at higher temperatures can explain the obser\'ation. Until we can work with pure enzymes, we shall have great difficulties in distinguishing a change in solubility from a true change in the enzyme.

- Enzymes and Reaction. Most enzymes are exactly adapted to the reaction of their "milieu" - that is to say, of the solution in which they occur in nature. Thus pepsin acts best in combination with acid, with hydrogen- ions. It does not digest protein, but only the chloride of the protein, and the products of the enzyme are the chlorides of peptones. The quantity of acid which is bound on the proteins and peptones, and which changes the free compounds into chlorides, is not suflScient for the action of pepsin. This was explained by Leo.^ Pepsin needs an excess of acid for its action, that is to say, it needs free hydrogen-ions. When we study dissolved protein, we cannot see this, because the hydrochloric- proteins split oflE acid by hydrolysis. But if we use a solid protein, for instance fibrin, we find that we must add so much hydrochloric acid, that the water, in which the swollen fibrin lies, still contains acid. The natural gastric juice of the dog or of man contains one-half or more per cent of hydrochloric acid, but the juice is diluted by food to 0.35 per cent; it is then i/io normal hydro- chloric acid, and it has been pointed out by Briicke^ and Pawlow, that the optimum of pepsin lies at this lower concentration. The hydrochloric acid can be replaced by another acid according to the available hydrogen-ions, but it seems that the hydrochloric acid has an exceptional value, which is greater than that of other acids of the same concentration and ionisation.

The necessity of co-operation between pepsin and hydrochloric acid has caused many errors in the chemistry of enzymes. The peptones produced by pepsin are basic, and neutralise the acids; the hydrochloric-peptones are hydrolysed salts, and the quantity of acid set free by hydrolysis depends upon the concentration and upon the presence of other neutral salts. All these influences appear to us as a diminution or an increase of pepsin. Furthermore, we know two proteins, the myosin of muscles or syntonin, and the fibrin of the blood, which swell in acids and can be attacked by pepsin only in the swollen state. Sodium chloride and other neutral salts check this swelling, and appear as opposed to pepsin; as '^antiferments." Pepsin is least suitable as an example

for pointing out the laws of enzymotic action.Un- fortunately, however, it is used most frequently. The relations of trypsin to the reaction have often been in- vestigated by earlier authors, but they did not always distinguish between the speed of action of the dissolved trypsin and the ease of dissolving trypsin from the gland.

The enzyme has two actions: it dissolves the protein, and it splits up proteins and peptones into simpler products. It also seems to have two different optima. It dissolves proteins when the reaction is distinctly alkaline; and it decomposes peptones best at the approximately neutral or slightly alkaline reaction which prevails in the intestine. This nearly neutral reaction is best also for the enzymes of the saliva, the intestine, and the tissues. For a long time there has been a division of opinion about the re- action of the contents of the intestine and the tissues.

The dispute arose solely because of the deficient knowledge at the time of the qualities of the indicators used in the study of the reaction. Now we know that the reaction of both tissues and contents of the intestine resembles that of a weak solution of alkali saturated with an excess of carbonic acid. Such a solution gives the optimum for the enzymes of the saliva and intestine, and for the oxidising enzymes of the tissues. The tissue enzymes are destroyed by an abnormal reaction, but the enzymes of the intestine show a greater or less resistance towards difi^erences in reaction, a remarkable adaptation to the conditions of their environment, because the reaction in the upper duodenum changes often and suddenly.

Zymase, the metabolic enzyme of the yeast, acts only, or best, in the presence of phosphates; of sodium mono- or diphosphate. Perhaps this is an example of a true activator, a matter which will be dealt with later on, but more probably the power of the phosphoric acid to hinder marked changes of reaction is the cause of the favourable influence of the phosphates, and this is therefore a fur- ther exemplification of the importance of reaction upon ferments.

- *Chemical Properties*: Enzymes combine both with their substrate and their dissociation products. The specificity of the enzymes mentioned before suggests to us, that the individual enzyiaes must have chemical relations with this substrate. We have no reason to think that all enzymes belong to the same class of chemical compounds. It has already been shown that the enzymes are accompanied as a rule by proteins and nucleic acids.

The difficulties of separating enzymes and proteins have led physiologists for a long time to regard enzymes as protein-like bodies, though Briicke * and others ^ were successful years ago in freeing enzymes from all traces of proteins. Perhaps steapsin is a fat-like compound, and sucroclastic enzymes sugar-like compounds, and perhaps they belong to a wholly different class. We know but little to-day about the chemical characteristics of enz)anes.

Enzymes are salted out by such neutral salts as are capable of readily salting out proteins, uric acid, and many dye-compounds, and particularly by anmionium sulphate. It seems that the individual enzymes are precipitated at diflferent concentrations of ammonium sulphate - ^aldehydase and erepsin, when the solution contains sixty per cent, and trypsin and invertin at complete saturation.

These diflferences can be used practically for purifying and separating enzymes, but they throw no light on their chemical nature. Enzymes are precipitated by alcohol, and the individual enzymes apparently by different concentrations of alcohol. En- zymes are precipitated, further, by heavy metals like uranium, and they are precipitated by tannin and by phosphotiingstic acid. The enzymes can be dissolved again immediately after precipitation, but on standing they readily lose their solubility. Colour tests for enzymes are not known.

It is of practical importance that enz)mies are neither precipitated nor destroyed by many antiseptic substances, *i.e.*, by substances that kill and destroy living cells or check their growth. Chloroform, toluene, thymol, and ether, which dissolve the lipoid substances in cells and bacteria, and therefore destroy structures necessary to life, exert no influence on enzymes, because the enz)anes are completely soluble in water, and do not yield any lipoids. It is very difficult to get extracts of living tissues, which cannot be heated or boiled, free from bacteria; so far as concerns the liver and pancreas, which yield bacteria in life, it is impossible.

During life, growth of micro-organisms is prevented or controlled in organs, but after death and loss of structure, bacteria grow in the fluids rich in proteins, carbohydrates, and suitable salts. These bacteria split the proteins, carbohydrates, and fats, as do the hydrolytic ferments, and give rise, like zymase, to carbon dioxide.

Every one who has worked with ferments knows how easily we can be deceived when led to suppose ferments to be present when in fact we are dealing with bacteria. We distin- guish between the two when we allow ferments to work for only two or three hours, because the action of ferments proceeds very quickly, whilst the growth of bacteria, even under the most favourable conditions, takes much more time.

But during extraction and purification, and if we extend experiments for a long time, it is absolutely necessary to protect enzyme solutions against bacteria by means of antiseptic substances. Amongst these chloroform or toluene have been most used within the last few years, and for the study of the enzymes of the alimentary canal crowded with bacteria, I suggest using both substances simultaneously. Thymol and ether have no efficiency.

I think that all observations on enzymes, made without any addition of antiseptics, must be regarded with doubt and distrust. Other antiseptics, which coagulate proteins, such as fluorides, potassium meta-arsenite, and mercuric chloride, seem to have an injurious effect on the action of the enzymes. But no evidence has been brought forward showing whether ferments are precipitated and destroyed by these compounds, or whether the proteins, on coagulat- ing, absorb and withdraw the enzymes. Buchner ^ has observed that zymase does not lose power in the presence of potassium meta-arsenite or ammonium fluoride in small amount, and that the proteolytic ferments of leucocytes are not harmed by corrosive sublimate.

But these solu- tions, zymase and the extracts of leucocytes, are rich in protein, and the harmful influence of the mercury or the arsenite is checked by their union with protein. Purified enzyme solutions have not yet been studied with regard to their sensitiveness to these antiseptics, but for most practical purposes the lipoid antiseptics are thoroughly satisfactory.

ACTIVE SITES AND CATALYTIC ACTION

TRIPLE ACTION CATALYSTS

Trifunctional organocatalysts that closely mimic natural enzymes can significantly increase reaction rates, say chemists in Japan. Tadashi Ema, Takashi Sakai and co-workers at University in Japan have made highly active organocatalysts by combining the reactivity of three separate organic components. Mimicking the catalytic triad structure found inside many enzyme active sites, the three components work cooperatively, accelerating a test reaction 3.7 million-fold.

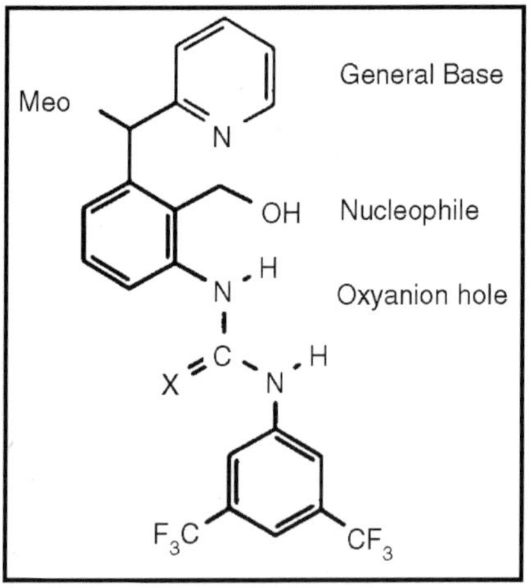

Fig. Catalytic Triad Structure of Many Enzymes Used as an Inspiration for an Organocatalyst

Ema's catalysts were inspired by lipase, an enzyme known to work through a catalytic triad. 'With precise information about the structures and reaction mechanism of lipases, we decided to test our abilities to mimic the active site,' said Ema. The resulting trifunctional organocatalyst has a nucleophilic OH group, a pyridine moiety, and either a urea or thiourea group. Ema tested his catalysts using the transesterification reaction, a synthetic transformation often applied to organic molecules to switch one ester side chain for another.

The catalyst's hydroxyl group, boosted in reactivity by the adjacent pyridine, initiates the reaction by attacking the ester group in vinyl trifluoroacetate - and the urea group stabilises the resulting intermediate through hydrogen bonding. Adding the catalyst increased the rate of reaction up to 3.7 million times, depending on the exact substrate used. For comparison, Ema also made three bifunctional control catalysts each missing a different one of the three components - and found all three to be essentially inactive.

'The observed rate accelerations are impressive,' said Ben List, who develops organocatalysts at the Institute. 'The authors have elegantly combined previously developed organocatalysis motifs to make a highly active and promising new organocatalyst type. Designing an asymmetric version is clearly the next logical step, which I look forward to.' Ema agrees that, along with further tuning the catalyst structure to boost reactivity levels even closer to those of enzymes, the next target will be to develop the catalyst for asymmetric synthesis.

ENZYMES AND INORGANIC CATALYSTS

Enzymes resemble inorganic catalysts in several aspects and differ from them in many other features. These similarities and differences are summarised in the following table.

Similarities	Diffferences
• Both are needed in imnute quantities	• Enzymes are complex proteins while catalysts are simple inorganic molecules
• Both accelerate the rate of a rection, but	• Enzymes have a very high weight while cannot initiate one molecular weight
• Both of them bring about a decrease in the activation energy	• Enzymes catalyse only biological reactions
• Bothe of them tempoarily combine with the substrate molecule	• Enzymes cataljyse only biological reactions
• Both of them do not undergo any chang in their composition	• Enzymes catalyse specific types kof reactions while cat-

Similarities	Differences
and hence, can be used again and agin	alysts have a wide range
• Both of them do not alter the nature and quantity of the end products	• Enzymes are regulated by specific substances called cofactors, but not so in the case of catalysts
• The reaction accelerated by both of them is reversible	• Enzymes can be easily inactivated but not so in the case of catalysts

14.12 - Similarities and Differences between Enzymes and Catalysts

Mode of Enzyme Action

There are two different views to explain the mode of enzyme action the lock and key hypothesis and the induced-fit hypothesis.

Lock and Key Hypothesis

It was suggested by Emil Fischer in 1894. According to this view, the enzyme molecule operates by chemically uniting with the substrate molecule, forming an enzyme-substrate complex. The enzyme molecule provides a uniquely structured template on which the substrate molecules can become attached and interact subsequently.

This brings about an interaction between the specific active sites in the enzyme molecule and the reactive sites in the substrate molecule. The enzyme now breaks down the substrate into products.

The products initially remain attached to the enzyme for a short while forming an enzyme product complex. The products get released from the enzyme molecule subsequently. The enzyme is now ready to receive another substrate molecule again. Thus, the same enzyme can be used again and again.

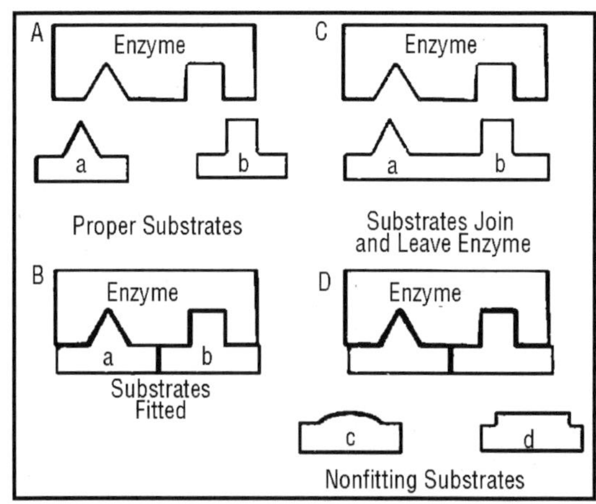

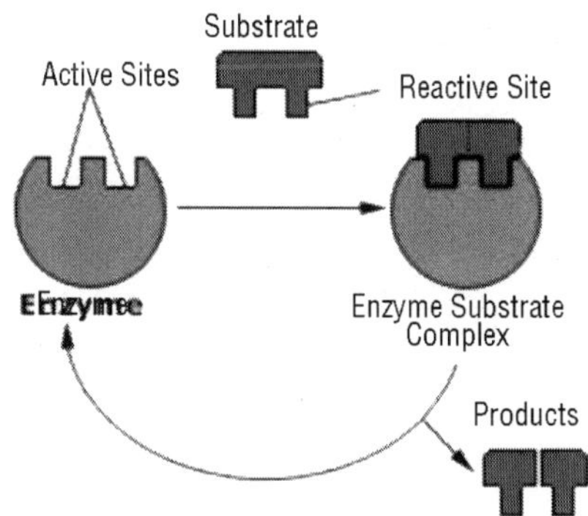

Fig. Breakdown Reaction Catalysed by an Enzyme According to Lock -and Key Hypothesis

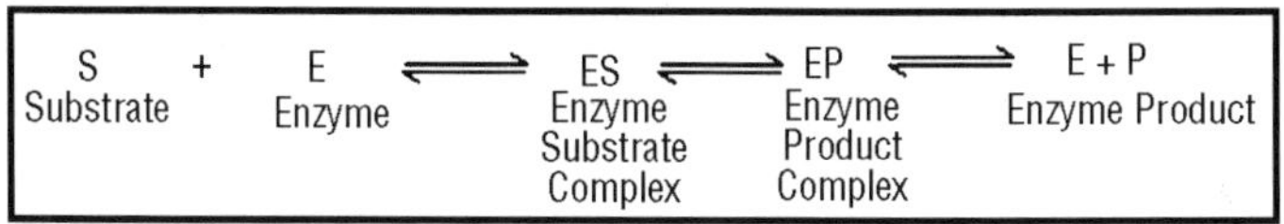

Fig. - Lock and Key Hypothesis

Induced Fit Hypothesis

It was proposed by Daniel Koshland in 1959. According to this hypothesis, there is an intermediate condition called transition state between the substrate and the products. It is highly unstable. When the substrate molecules bind to the enzyme molecule, a change is brought about in the active site to precisely fit the transition state (induced fit). This induced fit hold the substrates at the correct angle for the reaction to take place. The fact that an active site may also have a conformation to fit the product helps in explaining the role of enzymes in catalysing reversible reactions.

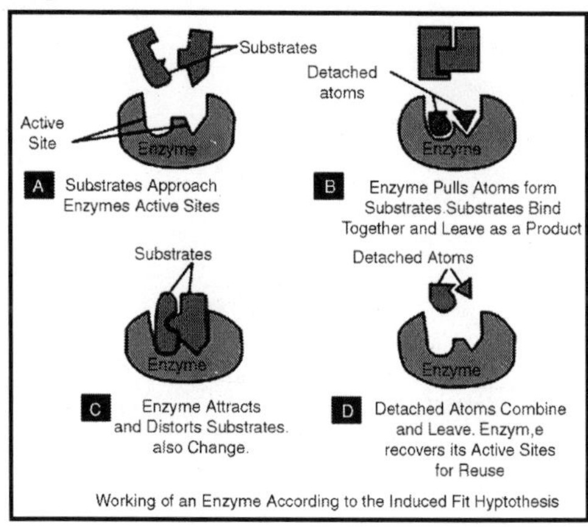

Fig. - Induced Fit Hypothesis

Isoenzymes

There are some enzymes which have slightly different molecular structure but exert similar catalytic action. Such enzymes are called isoenzymes or isozymes. More than 100 isozymes have been identified. The enzyme lactic dehydrogenase (LDH) in human skeletal muscle has five isozymes.

Subunit Association

Many enzymes can exhibit their catalytic activity only when they are in association with some non-protein substance. Such substances are called cofactors. The cofactors may be simple metal ions like Mg^{++} or complete organic compounds. An enzyme which functions only in the presence of a cofactor, is called apoenzyme. If the cofactor is an organic compound which can be easily separated from the apoenzyme, it is called coenzyme. If the cofactor is firmly bound with the enzyme, it is called prosthetic group. A working combination of an apoenzyme and its cofactor is called as holoenzyme.

CLASSIFICATION AND NOMENCLATURE

Numerous means exist for classifying and naming rocks, all of which arise because as humans we feel we have to recognise and categorise common or contrasting features in related things. Systems of nomenclature and classification may reflect: genetic, textural, chemical or mineralogical features.

GENTETIC

Basic system which classifies rocks on the basis of where they form.

Plutonic: At depth

Hypabyssal: Intermediate depth

Uolcanic: On the Earth's surface.

This system is not very practical, but it serves as a first approximation, it tells nothing about mineralogy, chemistry of the rocks and can not distinguish basalt from rhyolite.

TEXTURAL

Relies on the grain size of individual minerals in the rock.

Aphanitic: fine grained < 1 mm

Phaneritic: medium grained 1 to 5 mm

Coarse grained (pegmatitic) > 5 mm

This system has the same shortcomings as a genetic classification, however specific textures present may aid in classification, *e.g.*, phenocryst, ophitic, coronas, but these are not indicative of a specific environment of formation or a specific lithology.

CHEMICAL

This type of classification requires a complete chemical analysis of the rock in order to pigeonhole a sample, and is not practical under field conditions where only a hand lens and hammer are available. A chemical classification system has been proposed for volcanic rocks and a comparable scheme for plutonic rocks is not available. This leaves us with a system based on mineralogy.

MINERALOGICAL

The one gaining application is the result of several years work by the IUGS Subcommission on the Classification of Igneous Rocks or Streckeissen Classification.

ENZYMES AS CATALYSERS

We know of chemical processes in whicli the reacting substances completely disappear, giving rise to a new body. But in other processes one chemical compound reacts with another compound and changes it without entering into the final reaction itself. For instance, the conversion of cane sugar is expressed by the equation:

$$C_{12}H_{22}O_{11} + H_2O + wHCl = CeHijOe + CeHizOe + wHCL$$

The hydrochloric acid causes the reaction, but before and after the reaction we find the same quantity of acid. A second case is the formation of ether from alcohol by sulphuric acid:

$$2C_2H_5OH + WSO_4H_2 = C_2H_5.. C_2H_5 + H_2O + WSO_4H_2$$

The sulphuric acid is the cause of the reaction, but neither the original alcohol nor the end-products of the reaction contain the sulphuric acid, and the most import- ant point is the fact that there is no quantitative relation between the hydrochloric acid and the sugar or between the sulphuric acid and the alcohol and ether. Chemists * apply the term "catalytic" to such reactions, and it seems that all enzymotic processes are catalytic reactions.

The evidence for their designation as catalysers is that we cannot find any relation between the quantity of enzyme and the quantity of matter decomposed by it, and that, ">^ after the reaction, we find the enzyme is not consumed. The last argument, however, is not completely proved, because the estimation of the enzymes is not so exact as to exclude the consumption of a portion.

The great discrepancy between the small quantity of enzymos and the enormous quantity of matter converted by enzymes, is astonishing, and has impressed ail in- vestigators. The best explanation seems to be the admission of an intercalated reaction assumed by chemists generally in the case of the formation of ether.

Thus the reaction would be expressed by two equations:

- $C_2H_5OH + SO.H\text{^} = C_2H_5$ (HSOJ + H_2O).
- C_2H_5 (HSO J + $C_2H_5OH = C_2H_5.. C_3H, + H_3SO$.

That is to say, the molecule of the sulphuric acid is combined first with one molecule of alcohol, then it combines with a second molecule of alcohol, and finally it is liberated and can attack a new molecule of alcohol. It is highly probable that enzymes are combined thus with the substrate they work upon and with the products they yield. The arguments for this supposition will be cited later on, but since I have spoken of the enzymes as cataly- zers, the theory of Ostwald and Bredig should also be mentioned. They have ascribed a new function to cata- lysers besides the quality mentioned. They define a catalyser as a substance that never causes a reaction, but only accelerates a reaction going on spontaneously but more slowly without the catalyser. They propose the theory that the proteins or carbohydrates are always slowly dissociated in watery solution, and that the enzymes hasten this dissociation.

They thought that through this theory they could avoid the difficulty presented by the fact that enzymes mediate in such extensive processes, extensive also as regards the energy involved. They argue that our body produces its whole energy through the enz)mies oxidising and burning the food, and that it is difficult to suppose that the great quantity of energy four large calories per granmie of glucose, and nine per granmie of fat - can be liberated by the small quantity of enzymes which are available. The argument of Ostwald and Bredig has been adopted by many investi- gators; in fact we may say that it has ruled the whole theory of enzymotic action in recent years, but it seems to me that it does not hold good, and for three reasons:

- It is not true that proteins or carbohydrates spon- taneously undergo a slow dissociation in solution. Such solutions have been preserved for seventeen years or more at room temperature, and, in the absence of micro-organ- isms, of acid, or of alkali, the substances have been found unchanged. We have no reasons for thinking that the time of obser\"ation is too short, and that we would find a dissociation in centuries.
- If we heat cane sugar with acid, it is decomposed in the same manner as by invertin, and we could here attribute the decomposition really to the H-ions, which neutral water contains in lowest concentration, and assume simply that the enzyme hastens it. But in other cases, the enzymes cause other reactions than the H-ions of the water would bring about. The dissociation of glucose produced by H- or OH-ions gives rise to lactic acid and humins, while the dissociation by zymase produces alcohol.
- Sugars and fats are oxidised and burned in the tissue, and set free much energy, but we do not know exactly to what extent this oxidation is caused by enzymes, and especially by one enzyme. It is possible that the structure of cells is needed for the complete combustion, and that two, three, or more enz3mies work consecutively and perhaps in separate localities. We know that

Z3miase is the only enzyme that gives off heat, and only a very small quantity of heat. The well-known enzyiaes, the hydrolytic ones, liberate no heat. The action of trypsin, was measured by Grafe ^ in Rubner's laboratory with a calorimeter of high sensitiveness. He found no rise of temperature.

The other enzymes have not been measured in this manner, so far as I know, but the heat of combustion of starch, the disaccharides, and of glucose, has been determined, and we can calculate that the differences lie within the errors of analysis. The building-up of protein from the amino-acids, or of the polysaccharides from the hexoses, as well as the accompan)dng dissociations, seems to be unconnected with any perceptible thermic process. It is of great importance for the understanding of enzymes, that the hydrolytic enzymes neither give out nor consume any considerable or perceptible quantities of heat. Opin- ions regarding energy have here no weight. We must see whether well-observed facts support the theory that enzymes are like inorganic catalysers, and only hasten reactions without provoking them. For all proteolytic enzymes no fact is known which gives any evidence for it, and the same is true of ptyalin, invertin, lactase, zymase, and the corresponding metabolism enzymes urease and nuclease.

On the other hand, evidence has been brought forward that esters like fats are dissociat- ed spontaneously by the ions of the water, and that steap- sins or lipases which split up fats and other esters, only hasten this process. Later on, in treating of the synthetical action of enzymes, I shall discuss the steapsins, or lipases, which are distinguished from all other enzymes, because they alone effect synthesis to any great extent.

I think that the capacity for acting synthetically is connected with the property of the steapsins to hasten only a slow spon- taneous action. The steapsin seems to have the proper- ties of catalysers in the sense of Ostwald and Bredig, and this property lies at the bottom of the synthetical action. According to the most important work of Pottevin,^ the relations between dissociations and synthetical action in steapsin are perfectly similar to the relations between alcohols, acids, and esters in watery solution without enzymes, because both reactions show an equilibrium- point, which is moved in the one or the other direction according to the temperature or the presence of more or less water. If ethyl alcohol is heated with acetic acid, the following reaction occurs:

$$CH_3COOH + C_2H_5OH = CH_3CO.O.C_3H_5 + HA$$

But if water and ethyl acetate are mixed, the opposite reaction takes place, as follows:

$$CH_3CO.O.C_2H_5 + H,0 = CH_3COOH + C_2H_5OH.$$

If these two substances, alcohol and acetic acid, are allowed to mix, both reactions must proceed simultaneously in the same solution until they reach an equilibrium-point, that is to say, the point at which both proceed at the same rate, and all change seems to be stopped. The position of the equilibrium-point depends upon the mass of the three reacting substances, water, acid, and alcohol, and it depends, further, upon the temperature of the solution. The equation is therefore expressed in this manner:

$$CH_3COOH + C_2H50H?=\hat{}CH_3CO.O.C_3H_5 + HA$$

The steapsin moves the equilibrium-point, and thus acts like a true catalyser. In concentrated solutions it hastens s)mthesis, but if we add water, it hastens disso- ciation. It is remarkable now, that the disaccharides, maltose and isomaltose, are glucosides according to E. Fischer, and are similar to ether in structure, and that he even has observed strong hydrochloric acid to have a synthetical influence on glucose. Compounds like isomaltose are thus formed. No evidence has been brought forward thus far regarding the slow dissociation of maltose in the absence of maltase, but perhaps we can expect this fact to be observed, and in that case maltase must be likened to steapsin and not to sucroclastic enzymes which attack the anhydrous polysaccharides like starch or cane sugar, and which exhibit no catalytic quality.

A second class of enzymes, which hasten only those reactions proceeding slowly without them, are the oxidase;: of the type of laccase. The more important metabolic enzymes, like zymase or lactacidase, I repeat emphatically, provoke conversions which do not occur at all in the absence of enzymes, and so do all proteolytic enzymes and the nucleases. Bredig, Henri, and others have compared the time velocity of reactions, which are produced by inorganic catalysers and by enzymes, and they have found that the curves resemble one another, but show some dififerences. I have already stated that it is impossible to study the laws of action, if we have only solutions so crowded with impurities as are our enzyme solutions. I have said, and may repeat again, that tnzyiaes are modified in all sorts of ways in solutions, and shall presently mention that the action of enzymes is checked by the products formed by their action, and occurring therefore in the solutions; while in the case of ether or cane sugar, the adds pro- voking reactions are not influenced by. substances appeing in the solution. Thus irregular curves result, which throw light upon the happenings in enzymotic processes.

We can only observe that the action of enz5anes begins immediately when they touch their substrate, and that the process, with suitable reaction and temperature, proceeds most rapidly. If we add saliva, containing ptyalin, to starch, reducing sugars can be detected as soon as the fluids are mixed. Fibrin is dissolved in gastric juice, containing pepsin, as rapidly as sugar dissolves in water. If we allow active trypsin to act upon casein or other easily digestible protein, tyrosine is split off in so short a time that it resembles a mere precipitation or crystallisation of the insoluble amino-acid. When study- ing arginase, Kossel wished to heat a liver extract with an enzyme in order to destroy arginase, and he set the tube in boiling water.

The short time before the extract became heated was sufficient for the arginase to act upon the arginine and to split off some urea. The glycolytic enzyme of muscles does not convert more sugar in twelve hours than in two. The quick beginning and the rapid progress is a special feature of enzymotic action. If I find a ferment that acts only after a long time, proceeds slowly, and forms end-products only in small quantity, I always conclude that either the enzyme is of no great importance, or is not present under suitable conditions.

ENZYME PHOSPHOFRUCTOKINASE CATALYSES

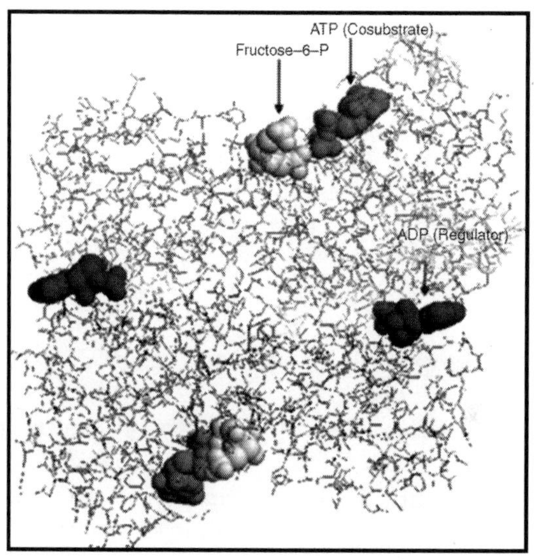

Fructose-6-phosphate + ATP '!
Fructose -1,6-bisphosphate + ADP

Structure of phosphofructokinase (wire-frame), with its substrates and the allosteric effector AMP bound to it (space-fill). In Figure the substrates are located next to each other within the active site of the dimeric enzyme. This reaction is an early step in the degradation of glucose, which ultimately serves to replenish ATP from ADP and phosphate.

It therefore makes good biological sense that phosphofructokinase should be stimulated by ADP. To accomplish this stimulation, ADP binds to another site that is far away from the active site; it therefore clearly does not directly participate in the reaction. Instead, ADP binding changes the conformation of the entire enzyme molecule. The change will also affect the active site and enhance the efficiency of catalysis there. This mode of action is known as allosteric regulation and is exceedingly common. Allostericeffectors can be either stimulatory (as ADP is in this example) or inhibitory. *E.g.*, ATP does not only bind to the active site but also acts as an allosteric inhibitor of phosphofructokinase.

The workings of allosteric regulation are schematically depicted Figure The enzyme has two possible conformations that are in equilibrium with each other An allosteric activator will bind selectively to the regulatory site as it occurs in the active conformation and thereby shift the equilibrium towards this conformation. Conversely, an inhibitor would bind selectively to the inactive conformation and thereby stabilise it. As you can see, activators and inhibitors may share the same regulatory site; this is the case in the above example of phosphofructokinase with ATP and ADP. Note, however, that phosphofrucokinase has additional allosteric sites that permit regulation by other effectors.

Enzyme regulation by allosteric effectors and by phosphorylation. a: The binding site for the allosteric effector (RS) is distinct from the active site (AS) of the enzyme.

The enzyme can assume two distinct conformations, and both the regulatory site and the active site are affected by the transition between them. Only in one conformation the enzyme is active. b: An inhibitor binds to a regulatory site in the inactive conformation, and it will change the conformational equilibrium accordingly. An activator stabilises the active conformation. c: Most allosteric enzymes are multimeric, and their transitions are synchronous. d: Regulation by phosphorylation also works by selective stabilisation of one conformation. Like allosteric regulation, it can be inhibitory (as shown here) or stimulatory.

Although it is not theoretically necessary, it seems that all allosteric enzymes occur as oligomers. In Figure dimeric enzyme is shown, but often the number of subunits is considerably higher. Their oligomeric nature enables enzymes to react more sensitively to changes in effector concentration. Regulation of phosphofructokinase and of fructose-1,6-bisphosphatase. a: Allosteric effectors have opposite effects on the two enzymes. b: Structure of fructose-2,6-bisphosphate (right). Fructose-1,6-bisphosphate is shown for comparison. c: Control of fructose-2,6-bisphosphate levels by hormones. See text for details.

Another important means of enzyme regulation consists in *phosphorylation*. This occurs by protein kinases, which transfer a phosphate group from ATP to a specific site on the regulated enzyme. The mechanism of regulation by phosphorylation is not really that different – it also works by selective stabilisation of either the active or the inactive conformation.

The only difference is that the regulator is more stably attached, so that the regulatory effect may last longer. Like allosteric regulation, it can be inhibitory (as shown in Figure or stimulatory. Many enzymes are subject to regulation both by allosteric effectors and by phosporylation.

While all mechanisms discussed so far modulate the activity of existing enzyme molecules, the overall enzyme activity may also be varied by changing their abundance:

- The transcription of the gene encoding the enzyme in question can be turned on or off. This mechanism is employed by many hormones, in particular steroid hormones (*e.g.*, cortisone) and thyroid hormones.
- The stability of the enzyme mRNA's of the enzymes can be changed.

Hormones may affect the activity of an enzyme at more than one level. For example, insulin increases the activity of glycogen synthase by way of transcriptional induction, increased mRNA stability, and protein phosphorylation.

ENZYMES

In such manner Kossel and Dakin * freed liver arginase from most proteins. Essentially the same principle applies when fresh tissue-extracts are al- lowed to stand for some time; coagulation being then due to he weak acid reaction resulting in muscles and other tissues shortly after death. If we dialyse the extract at this time in running water, we remove both the proteins which become insoluble and the salts and other crystalloid substances which pass through the parchment. After the dialysis, or after standing, we filter oflE the precipitate.

As a rule it is useless to filter under pressure through a Chamberland or similar filter. Nevertheless, the advantage of Buchner's method con- sists not only in the complete disintegration of tissues and structures, but likewise in the filtration of the press- juice through siliceous earth, which retains the nucleoproteins, the globulins, and all proteins coagulating spontaneously. Almost the only proteins found in the press-juice of the muscles and the liver are albumin and the haemoglobin of the blood. These juices are in most cases very rich in enzymes. Using one of these methods it is easy to obtain proteolytic enzymes and steapsins, but many sucroclastic enzymes are retained by the proteins, and cannot be dissolved out again by water. Sometimes it is possible to dissolve them by very weak alkali, but the necessary conditions have not been suflSciently studied.

If we do not succeed in bringing the enzymes into solution, we must add the solid precipitate to the solution containing carbohydrates or fats or proteins, which are to be acted upon by the enzyme. We do not know whether the enzyme works under such conditions in a solid state and converts carbohydrates and other substances only by contact, or whether the enzyme passes into solution.

We shall see that we must assume a chemical connection and combination between enzyme and substance. Per- haps the chemical compound,thus newly formed is more soluble than the free enzyme. This would be a good ex- planation of the action, for instance, by which the mucous membrane of the small intestine converts milk sugar, although it is impossible to find lactase in the extracts, or the ability of dried pancreas powder to dissolve protein,which is lacking in extracts made from this powder. We must always remember, however, that the difference can be produced also by an improper reaction of the extract. It is an important fact, however, that we can use the dry powder for many purposes.

With these methods we can easily remove proteins and other harmless, indifferent substances. The chief diffi- culties which complicate the purification, extraction, and investigation of enzymes, are the presence of other enzymes, and the anchorage of the enzymes in the cells. The avoidance of these difficulties is the chief advantage of the secreted natural juices as compared with the extracts.

Another without great loss. The gastric juice contains only the well-known proteolytic enzyme, pepsin, which dissolves natural proteins and converts them into peptones; but the process never goes beyond the stage of yielding biuret-giving substances. The mucous membrane of the stomach, and therefore its extracts, contains, besides pepsin destined for secretion, an erepsin acting intracellularly and splitting peptones into amino-acids. The pepsin and the erepsin have the same solubilities, and if we free the proteolytic enzyme from impurities, we concentrate both enzymes, getting an artificial juice with properties other than those of the natural one. The con- tradictions of authors on the question of trypsin and trypsinogen is believed by Bayliss * to be susceptible of the same explanation.

If we microscopically study the pancreas or a salivary gland at rest, we see the cell crowded with small, refractive granules forming a mass that leaves only a very narrow clear space next to the basement membrane. In the salivary gland the cell can be wholly studded with the granular mass. During glandular activity the granfiles become much fewer in number, and retract to an inner narrow margin; in the salivary gland they may even dis- appear. There can be no doubt that during the act of secretion, the granules are discharged to form part of the secretion. They become the solid matter of the juices; therefore they must be soluble in water. If we extract the pancreas or the parotid gland while studded with these granules, they, and with them the enzymes, are dissolved immediately. As has been pointed out by Pekelharing,^ we find in the extracts of the gastric mucous membrane the most characteristic chemical compound of the gastric juice.

There is something else to be noticed in dealing with the intracellular enzymes that never leave the cell. In the state of activity, both these enzymes and the substances hydrolysed or oxidised by them must be in solution, for "corpora non- agunt nisi soluta." To avoid the action of enz)anes at rest, the two bodies must be separated from each other; they must be fixed outside the real fluid protoplasm. We know another substance which, in this respect, is analogous to the endo-enzymes; glycogen, which is soluble in water, but which may be stored in insoluble form within the cell. Ehrlich ^ found that glycogen is fixed in the liver cell by a "Trager- substanz," a holding-substance, and that one of the colour reactions attributed to glycogen is due to this substance. Glycogen and this Tragersubstanz are rather firmly bound together. We know with what difficulty the glycogen is completely dissolved out of the organs, and with what difficulty it is freed from the last traces of impurities. It seems to me that we must also assume such a holding-substance for the intracellular enzymes which restrains the action of the enzyme even after the almost complete disintegration of the tissues.

3

Bioprocess Parametres in Enzyme Production

Any operation involving the transformation of some raw material (biological or non-biological) into some product by means of microorganisms, animal or plant cell cultures, or by materials derived from them (*e.g.* enzymes, organelles), may be termed as a "bioprocess". Until about 1950, the predominant method of producing industrial enzymes was by extraction from animal or plant sources; by 1993, this accounts for less than 10%. Industrial enzymes are now produced by microorganisms grown in aqueous suspension in large vessels, *i.e.*, by fermentation.

Microorganisms are preferred to plants and animals as sources of enzymes because:

- They are generally cheaper to produce;
- Their enzyme contents are more predictable and controllable;
- Reliable supplies of raw material of constant composition are more easily arranged;
- Plant and animal tissues contain more potentially harmful materials, including phenolic compounds (from plants) and endogenous enzyme inhibitors.

Enzymes are usually sensitive to harsh physical and chemical conditions, which demands careful selection of production processes and conditions for each individual enzyme. In aerobic bioprocesses, the important criteria that must be taken into account, in order to have high product yield, are:

- Microorganism
- Medium composition
- Bioreactor operation parametres
  - Temperature
  - pH
  - Oxygen transfer rate
    a. Air inlet rate (Q0/VR)
    b. Agitation rate (N)

MICROORGANISM

In bioprocesses, the selection of host microorganism for production of industrial enzymes is often critical for the commercial success of the product. Potential hosts should give sufficient yields, be able to secrete large amounts of protein, be suitable for industrial fermentations, produce a large cell mass per volume quickly and on cheap media, be considered safe based on historical experience or evaluation by regulatory authorities, and should not produce harmful substances or any other undesirable products. Few microorganisms fulfill all the above criteria and are used for production of industrial enzymes. *Escherichia coli* is one of the most important host microorganisms. Benzaldehyde lyase is naturally produced by wild-type *Pseudomonas fluerescens*. Hinrichsen cloned the gene encoding benzaldehyde lyase to an *Escherichia coli* strain; the studies until then has been performed using this host microorganism.

Escherichia coli

If posttranslational modifications are unnecessary, *E. coli* is most often chosen as the initial host. The main reason for the popularity of *E. coli* is the broad knowledge base for it. *E. coli* physiology and its genetics are probably far better understood than for any other living organism. This large knowledge base greatly facilitates sophisticated genetic manipulations. Also, an important engineering contribution was the development of strategies to grow cultures of *E. coli* to high cell densities.

The average *E. coli* cell is a rod shaped body with diametre and length of approximately 1 and 2 micrometres, respectively. It has a wet mass of 9.5×10^{-13} g, or 2.9×10^{-13} g dry mass. Elemental essay of the dry mass of *Escherichia coli* cells reveals a fairly typical composition of the protoplasm: 50% carbon, 20% oxygen, 14% nitrogen, 8% hydrogen, 3% potassium, 2% sodium, 1% sulfur, 0.05% calcium, 0.05% magnesium, 0.05% chlorine, 0.2% iron and a total of 0.3% trace elements including manganese, cobalt, copper, zinc and molybdenium. *E. coli* is a Gram-negative bacterium. The murein layer of the Gramnegative bacteria is much thinner than that of Gram-positive bacteria, and they make a completely different structure- an outer membrane; which has the ability to resist damaging chemicals. *E. coli* is a facultative aerobe bacterium which can grow in the presence or absence of oxygen. Under anaerobic conditions, *E. coli* can utilize NO_3 as a source of nitrogen, but when growing aerobically it requires a reduced nitrogen source.

The maximum temperature of growth for *E. coli* is approximately 48°C, its optimal temperature is 37°C, its minimum temperature is 8°C, and its normal temperature range extends from about 21°C to 39°C. In the absence of methionine, growth stops at 45°C. *E. coli* grows well near neutrality at pH values from 6.0 to 8.0. Regardless of the value of the external pH, within the cell, pH is maintained at a value quite close to 7.6.

Cell Growth, Kinetics and Yield Factors

For microorganisms, growth is their most essential response to their physiochemical environment. Growth is a result of both replication and change in cell size. In a suitable nutrient medium, organisms extract nutrients from the medium and convert them into biological compounds. Parts of these nutrients are used for energy production and parts are used for biosynthesis and product formation. Microbial growth is a good example of an autocatalytic reaction. The rate of growth is directly related to cell concentration, and cellular reproduction is the normal outcome of this reaction. The rate of microbial growth is characterized by the specific growth rate, ì, which is defined as,

$$\mu = \frac{1}{C_X} \cdot \frac{dC_X}{dt}$$

where CX is the cell mass concentration (kg m^{-3}), t is time (h), and ì is the specific growth rate (h-1). When a liquid nutrient medium is inoculated with a seed culture, the organisms selectively take up dissolved nutrients from the medium and convert them into biomass.

A typical batch growth curve includes the following phases:

- Lag phase; occurs immediately after inoculation and is a period of adaptation of cells to a new environment ($\mu \approx 0$).
- Exponential growth phase; the cells have adjusted to their new environment. Growth achieves its maximum rate ($\mu \approx \mu_{max}$).
- Deceleration phase; growth decelerates due to either depletion of one or more essential nutrients or the accumulation of toxic by-products of growth ($\mu < \mu_{max}$).
- Stationary phase; the net growth rate is zero (no cell division) or the growth rate is equal to the death rate ($\mu = 0$).
- Death phase; cells lose viability and lyse ($\mu < 0$).

Despite the complexity occurring in cell growth, yield principles can be applied to cell metabolism to relate flow of substrate in metabolic pathways to formation of biomass and other products. Yield coefficients allow to quantify the nutrient requirements and production characteristics of an organism. Yield coefficients are defined based on the amount of consumption of another material.

For example, the growth yield in a fermentation is:

$$Y_{x/s} = \frac{\Delta X}{\Delta S}$$

where, $Y_{X/S}$ is the yield coefficient, X and S are mass of cell and substrate, respectively, involved in metabolism. This definition gives an overall yield representing some sort of average value for the entire culture period. However, in batch processes, the yield coefficients may show variations throughout the process for a given microorganism in a given medium, due to the growth rate and metabolic functions of the microorganism. Therefore, it is sometimes necessary to evaluate the instantaneous yield at a particular point in time. Instantaneous yield can be calculated as follows:

$$Y_{x/s} = \frac{dx}{ds} = \frac{dX/dt}{dS/dt} = \frac{r_X}{r_S}$$

Yield coefficients based on other substrates or product formation may be defined; for example,

$$Y_{x/o} = \frac{\Delta x}{\Delta O}$$

$$Y_{P/S} = \frac{\Delta P}{\Delta S}$$

For organisms growing aerobically on glucose, YX/S is typically 0.4 to 0.6 g/g for most yeast and bacteria, while YX/O is 0.9 to 1.4 g/g. Anaerobic growth is less efficient, and the yield coefficient reduced substantially.

IMMOBILIZED ENZYMES IN BIOPROCESS

Immobilization of biocatalysts helps in their economic reuse and in the development of continuous bioprocesses. Biocatalysts can be immobilized either using the isolated enzymes or the whole cells. Immobilization often stabilizes structure of the enzymes, thereby allowing their applications even under harsh environmental conditions of pH, temperature and organic solvents, and thus enable their uses at high temperatures in nonaqueous enzymology, and in the fabrication of biosensor probes. In the future, development of techniques for the immobilization of multienzymes along with cofactor regeneration and retention system can be gainfully exploited in developing biochemical processes involving complex chemical conversions. The present review outlines some of the above aspects, and delineates the present status and future potentials of immobilized enzymes and nonviable cells in the emerging biotech industries.

Biotechnology is currently considered as a useful alternative to conventional process technology in industrial and analytical fields. This is mainly because, unlike the chemical catalysts, the biological systems have the advantages of accomplishing complex chemical conversions under mild environmental conditions with high specificity and efficiency. Biological systems help in ingredient substitution, processing aid substitution, more efficient processing, less undesirable products, increased plant capacity, increased product yields, and improved or unique products. The variety of chemical transformations catalysed by enzymes has made these catalysts a prime target of exploitation by the emerging biotech industries. Despite these advantages, the use of enzymes in industrial applications has been limited by several factors, mainly the high cost of the enzymes, their instability, and availability in small amounts. Also the enzymes are soluble in aqueous media and it is difficult and expensive to recover them from reactor effluents at the end of the catalytic process. This restricts the use of soluble enzymes to batch operations, followed by disposal of the spent enzyme-containing solvent. Over the last few decades, intense research in the area of enzyme technology has provided many approaches that facilitate their practical applications. Among them, the newer technological developments in the field of immobilized biocatalysts can offer the possibility of a wider and more economical exploitation of biocatalysts in industry, waste treatment, medicine, and in the development of bioprocess monitoring devices like the biosensor.

Immobilization means associating the biocatalysts with an insoluble matrix, so that it can be retained in a proper reactor geometry for its economic reuse under stabilized conditions. Immobilization thus allows, by essence, to decouple the enzyme location from the flow of the liquid carrying the reagents and products. Immobilization helps in the development of continuous processes allowing more economic organization of the operations, automation, decrease of labour, and investment/capacity ratio. Immobilized biocatalysts offer several other advantages, notable among them is the availability of the product in greater purity. Purity of the product is very crucial in food processing and pharmaceutical industry since contamination could cause serious toxicological, sensory, or immunological problems. The other major advantages include greater control over enzymatic reaction as well as high volumetric productivity with lower residence time, which are of great significance in the food industry, specially in the treatment of perishable commodities as well as in other applications involving labile substrates, intermediates or products·

IMMOBILIZED NONVIABLE CELLS AS AN ECONOMICAL SOURCE OF ENZYMES

Biocatalysts can be immobilized using either the isolated enzymes or the whole cells or cellular organelles. Immobilization of whole cells has been shown to be a better alternative to immobilization of isolated enzymes· Doing so avoids the lengthy and expensive operations of enzyme purification, preserves the enzyme in its natural environment thus protecting it from inactivation either during immobilization or its subsequent use in continuous system. It may also provide a multipurpose catalyst, specially when the process requires the participation of number of enzymes in sequence. The major limitations which may need to be addressed while using such cells are the diffusion of substrate and products through the cell wall, and unwanted side reactions due to the presence of other enzymes. The cells can be immobilized either in a viable or a nonviable form. Immobilized nonviable cell preparations, which are normally obtained by permeabilizing the intact cells, for the expression of intracellular activity are useful for simple processes that require single-enzyme with no requirement for cofactor regeneration, like hydrolysis of sucrose or lactose· On the other hand immobilized viable cells, which serve as 'controlled catalytic biomass', have opened new avenues for continuous fermentation on heterogeneous catalysis basis by serving as self-proliferating biocatalysts· The present review will focus mainly on immobilized isolated enzyme systems, and the use of immobilized nonviable cells as a source of enzymes. For the aspects on the immobilized viable cell systems, the readers can refer to another article in this issue. Some of the aspects on the applications of immobilized viable cells in biochemical process development and monitoring have also been reviewed recently·

Most of the enzymes used at industrial scale are normally the extracellular enzymes produced by the microbes. This has been mainly due to their ease of isolation as crude enzymes from the fermentation broth. Moreover, the extracellular enzymes are more stable to external environmental perturbations compared to the intracellular enzymes. However, over 90% of the enzymes produced by a cell are intracellular. The economic exploitation of these, having a variety of biochemical potentials, has been limited in view of the high cost involved in their isolation. Also, compared to extracellular enzymes, the intracellular enzymes are more labile. Delicate and expensive separation methods are required to release the enzymes undamaged from the cell, and to isolate them. This increases the labour and the cost of the enzyme. These problems could now be obviated by the use of permeabilized cells as a source of enzyme. Permeabilization of the cells removes the barrier for the free diffusion of the substrate/product across the cell membrane, and also empties the cell of most of the small molecular weight cofactors, etc., thus minimizing the unwanted side reactions. Side reactions, which can occur due to the presence of other enzymes in a cell, can also be minimized by inactivating such enzymes prior to or after immobilization· Such permeabilized cells, which are often referred to as nonviable or nongrowing cells, can be exploited in an immobilized form as a very economical source of intracellular enzyme for simple bioconversions like hydrolysis, isomerization and oxidation reactions that do not need a cofactor-regeneration system· The decision to immobilize cells either in a viable or nonviable form is very important and depends on their ultimate application. Thus the permeabilized *K. fragilis* cells, which are nonviable, convert lactose in milk to glucose and galactose· whereas the viable (nonpermeabilized) cells convert the lactose to ethanol and are useful in the complete desugaration of milk·

A variety of physical, chemical, and enzymatic techniques have been developed for the permeabilization of cells· Some techniques like entrapment in polyacrylamide itself have been shown to result in permeabilization of the cells· A few typical examples of immobilized enzyme preparation obtained using permeabilized cells.

TECHNIQUES AND SUPPORTS FOR IMMOBILIZATION

A large number of techniques and supports are now available for the immobilization of enzymes or cells on a variety of natural and synthetic supports. The choice of the support as well as the technique depends on the nature of the enzyme, nature of the substrate and its ultimate application. Therefore, it will not be possible to suggest any universal means of immobilization. It can only be said that the search must continue for matrices which provide facile, secure immobilization with good interaction with substrates, and which conform in shape, size, density and so on to the use for which they are intended. Care has to be taken to select the support materials as well as the reagents used for immobilization, which have GRAS status, particularly when their ultimate applications are in the food processing and pharmaceutical industries.

Macromolecular, colloidal, viscous, sticky, dense or particulate food constituents or waste streams also limit the choice of reactor and support geometries. Commercial success has been achieved when support materials have been chosen for their flow properties, low cost, nontoxicity, maximum biocatalysts loading while retaining desirable flow characteristics, operational durability, ease of availability, and ease of immobilization· The variety of techniques and supports investigated for the immobilization have been reviewed in a number of articles and books· The objective of the preceding part of the paper is to review briefly, in the light of current developments, some approaches that have been used for the immobilization of biocatalysts with reference to the studies carried out in the author's laboratory as well as few others.

Techniques for immobilization have been broadly classified into four categories, namely entrapment, covalent binding, cross-linking and adsorption. A combination of one or more of these techniques has also been investigated. It must be emphasized that in terms of economy of a process, both the activity and the operational stability of the biocatalysts are important. They determine its productivity, which is the activity integrated over the operational time.

IMMOBILIZATION OF CELLS

Entrapment has been extensively used for the immobilization of cells, but not for enzymes. The major limitation of this technique for the immobilization of enzymes is the possible slow leakage during continuous use in view of the small molecular size compared to the cells. Biocatalysts have been entrapped in natural polymers like agar, agarose and gelatine through thermoreversal polymerization, but in alginate and carrageenan by ionotropic gelation· A number of synthetic polymers have also been investigated. Notable among them are the photo-crosslinkable resins, polyurethane prepolymers, and acrylic polymers like polyacrylamide·

Among these, the most widespread matrix made from monomeric precursors is the polyacrylamide gel. Polyacrylamide may not be a useful support for use in food industry in view of its toxicity, but can have potentials in the treatment of waste and in the fabrication of analytical devices containing biocatalysts. One of the major limitations of entrapment technique is the diffusional limitation as well as the steric hindrance, especially when the macromolecular substrates like starch and proteins are used. Diffusional problems can be minimized by entrapment in fine fibres of cellulose acetate or other synthetic materials or by using an open pore matrix· Recently, the development of so-called hydrogels and thermoreactive water-soluble polymers, like the albumin-poly (ethylene glycol) hydrogel, have attracted attention in the field of biotechnology· In the area of health care, they offer new avenues for enzyme immobilization. Such gels with a water content of about 96% provide a microenvironment for the immobilized enzyme close to that of the soluble enzyme with minimal diffusional restrictions·

THE IMMOBILIZATION OF ENZYMES

Covalent binding is an extensively used technique for the immobilization of enzymes, though it is not a good technique for the immobilization of cells. The functional groups extensively investigated are the amino, carboxyl, and the phenolic group of tyrosine· Enzymes are covalently linked to the support through the functional groups in the enzymes, which are not essential for the catalytic activity. It is often advisable to carry out the immobilization in the presence of its substrate or a competitive inhibitor so as to protect the active site. The covalent binding should also be optimized so as not to alter its conformational flexibility. Some of these problems however, can be obviated by covalent bonding through the carbohydrate moiety when a glycoprotein is concerned. A number of industrially useful enzymes are glycoproteins wherein the carbohydrate moiety may not be essential for its activity. In general, functional aldehyde group can be introduced in a glycoprotein by oxidizing the carbohydrate moiety by periodate oxidation without significantly affecting the enzyme activity. The enzyme could then be covalently linked to a support containing an alky amine group through Schiffs base reaction. Enzymes like glucose oxidase, peroxidase, invertase, etc. have been immobilized using this technique·

Covalent binding has been extensively investigated using inorganic supports. Enzymes covalently bound to inorganic supports have been used in the industry· Enzymes have also been bound to synthetic membranes, thus integrating biconversion and downstream processing· Large-scale processes using such an approach have been demonstrated for the preparation of invert sugar using invertase·

Iimmobilized through chemical cross-linking

Biocatalysts can also be immobilized through chemical cross-linking using homo- as well as heterobifunctional cross-linking agents· Among these, glutaraldehyde which interacts with the amino groups through a base reaction has been extensively used in view of its GRAS status, low cost, high efficiency, and stability· The enzymes or the cells have been normally cross-linked in the presence of an inert protein like gelatine, albumin, and collagen· Studies from our laboratory have shown the use of raw hen egg white as an economic, easily available novel proteinic support rich in lysozyme for the immobilization of enzyme or nonviable cells either in a powder, bead or highly porous foam form, using glutaraldehyde as the cross-linker.

The unique feature of this support is the large concentration of lysozyme naturally present in hen egg white which gets co-immobilized, thus imparting the bacteriolytic property to the support· Adsorption followed by cross-linking has also been used for the immobilization of enzymes. The technique of cross-linking in the presence of an inert protein can be applied to either enzymes or cells. The technique can also be used for the immobilization of enzymes by cross-linking the cell homogenates· Osmotic stabilization of cellular organelles or halophilic cells prior to immobilization using cross-linkers has also shown promise.

Adsorption

This is perhaps the simplest of all the techniques and one which does not grossly alter the activity of the bound enzyme. In case of enzymes immobilized through ionic interactions, adsorption and desorption of the enzyme depends on the basicity of the ion exchanger. Moreover, a dynamic equilibrium is normally observed between the adsorbed enzyme and the support which is often affected by pH as well as the ionic strength of the surrounding medium. This property of reversibility of binding has often been used for the economic recovery of the support. This has been successfully adapted in industry for the resolution of racemic mixtures of amino acids, using amino acid acylase· A variety of commercially available ion exchangers have been investigated for this purpose· One of the techniques, which has gained importance more recently, is the use of polyethylenimine for imparting polycationic characteristics to many of the neutral supports based on cellulose or inorganic materials· Enzymes with low pI, like invertase, urease, glucose oxidase, catalase, and other enzymes have been bound through adsorption followed by cross-linking on polyethylenimine-coated supports.

Immobilization of enzymes through hydrophobic interaction has also shown promise· One of the important features of this technique, which is of great significance, is that, unlike ionic binding, hydrophobic interactions are usually stabilized by high ionic concentrations, thus enabling the use of high concentrations of substrates as desired in an industrial process without the fear of desorption. Other types of strong interactive binding techniques have also been reported for the reversible immobilization of enzymes.

A typical example is the immobilization of soybean b -amylase on phenyl-boronate agarose, which can be reversed for the recovery of the support using sorbitol· Varieties of biospecific interactions have also been investigated for the reversible immobilization of enzymes by adsorption. Enzymes like acetyl choline esterase, ascorbic acid oxidase, invertase, peroxidase, glucose oxidase, etc. have been immobilized by biospecific-reversible immobilization on lectin-bound supports and invertase, using polyclonal antiinvertase antibodies· Traditional enzyme immobilization procedures involve isolation of the enzyme, followed by use of several steps for the immobilization. Costs of enzyme purification, the immobilization procedure, bioreactor operational stability, and bioreactor regeneration are the major factors that determine the cost of a bioreactor process. Development of techniques for the simultaneous isolation and immobilization of enzymes from crude extracts has obviated these problems. Some typical examples include the immobilization of a streptavidin-b -galactosidase fusion protein expressed in *Escherichia coli* and bioselectively adsorbed from a crude cell lysate to biotin which is covalently immobilized on controlled pore glass' and the simultaneous purification and immobilization of D-amino acid oxidase from *Trigonopsis variabilis* cell lysate adsorbed on phenyl sepharose· The technique can have future potentials especially in the downstream processing and immobilization of enzyme/proteins obtained by recombinant DNA technology.

Techniques for the adhesion of whole cells on polymeric surfaces are also currently gaining considerable importance. The major advantage for the cells immobilized through adhesion is reduction or elimination of the mass transfer problems associated with the commonly used gel entrapment method. The technique of immobilization usually being followed is the microbial colonization by recycling of the cell suspension along with nutrients, such that a biofilm is gradually formed. This often results in the immobilization of cells in a viable form for use in heterogeneous fermentations·

Useful techniques have been developed also for the immobilization of nonviable cells to be used as an enzyme source for simple chemical conversions. Notable among them include treating the cells or the support with trivalent metal ions like Al or Fe^{3+} or charged colloidal particles' and use of polycationic polymers like chitosan. Novel techniques have been developed to adhere cells strongly on a variety of polymeric surfaces including glass' cotton cloth' cotton threads' and other synthetic and inorganic surfaces using polyethylenimine. This technique has also been recently used for the simultaneous filtration and immobilization of cells from a flowing suspension, thus integrating downstream processing with bioprocessing. This technique may have future potentials for the immobilization of cells for food applications. Polyethylenimine is nontoxic and the United States of America Food and Drug Administration has permitted its use as a direct food additive under the Food Drug and Cosmetic Act.

THE SPECIFICITY AND MECHANISM OF ENZYME ACTION

The action of enzymes, unlike that of inorganic catalysts, is strictly specific and depends on the structure of the substrate on which the enzyme acts. A good example of this dependence is the catalyzed reaction of the hydrolytic breakdown of the amino acid arginine into ornithine and urea by arginase:

$$\underset{\text{Arginine}}{(H_2N)(HN{=})C{-}NH{-}CH_2{-}CH_2{-}CH_2{-}CH(NH_2)COOH} + H_2O \xrightarrow{\text{Arginase}}$$

$$\rightarrow \underset{\text{Urea}}{CO(NH_2)_2} + \underset{\text{Ornithine}}{H_2N{-}CH_2{-}CH_2{-}CH_2CH(NH_2)COOH}$$

However, arginase does not decompose the methyl ether of arginine:

$$(H_2N)(HN{=})C{-}NH{-}(CH_2)_3{-}CH(NH_2)COOCH_3$$

A dipeptide consisting of the radicals of two molecules of arginine yields only half of the theoretical quantity of urea when the dipeptide is acted on by arginase. It is evident that although the decomposition of arginine occurs in an area remote from the carboxyl group COOH (indicated by the broken line), an essential condition for the action of arginase is its combination with the carboxyl group of arginine. For this reason, either displacement of the hydrogen in the carboxyl group to the methyl radical or the bonding of the carboxyl group with the second molecule of arginine strongly influences the action of arginase.

Examples of the specificity of enzyme action may be observed by examining the stereochemical specificity of enzymes, that is, the action of enzymes on stereoisomers. For example, enzymes that oxidize natural L-amino acids do not affect the D-isomers of the same amino acids, and the enzyme dipeptidase, which hydrolyzes dipeptides consisting of radicals of L-amino acids, does not affect dipeptides composed of D-amino acid residues. Observation of the specificity of enzyme action led the German chemist E. Fischer to compare a given substrate and the enzyme catalyzing its transformation to a lock and its key. The stereochemical specificity of enzymes is closely related to one of the basic features of living organisms: their ability to synthesize optically active organic compounds.

Only those functional groups of an enzyme molecule that constitute the molecule's active centre take part in the union between the enzyme and the substrate, a union termed the enzyme-substrate complex. For example, in a molecule of chymotrypsin, an enzyme that hydrolizes proteins and that consists of 246 amino-acid radicals, the active centre is composed of one of the residues of serine (chymotrypsin is one of the serine proteinases) and two radicals of histidine located in sections of the polypeptide chain that are remote from one another.

These functional groups of the active centre converge because of the specific spatial (tertiary) structure that is typical of the chymotrypsin molecule. The disruption of this structure owing to protein denaturation or to other types of chemical modification leads to the alteration or cessation of catalytic activity. In compound enzymes, both the prosthetic group and the functional groups of an apoenzyme take part in forming the enzyme-substrate complex. For example, in the decomposition of pyruvic acid by pyruvate decarboxylase, the substrate is bonded to part of a thiamine pyrophosphate molecule in the following manner:

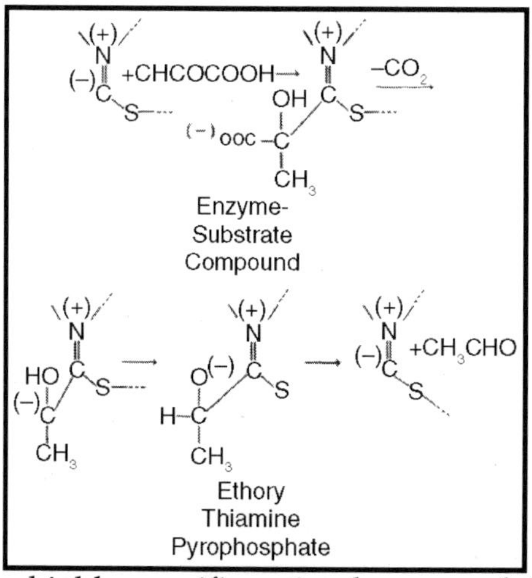

+CHCOCOOH → $-CO_2$

Enzyme-
Substrate
Compound

$+CH_3CHO$

Ethory
Thiamine
Pyrophosphate

Enzymes have a highly specific action because of their protein nature. For example, pyridoxal enzymes that contain the coenzyme codecarboxylase may belong to different classes and may catalyze diverse reactions. The specificity of these enzymes' action depends on the nature of the given apoenzyme.

Conditions of Enzyme Action

Enzyme action depends on a number of factors, primarily temperature and the reaction of the medium (pH). The temperature conducive to optimum enzymatic activity is generally in the range of 40°–50°C. At lower temperatures the enzymatic reaction is generally slower, and at temperatures approaching 0°C, it virtually ceases. Temperatures higher than 40°–50°C also slow down the enzymatic reaction and eventually inactivate it. The reduced intensity of enzyme action in the presence of higher temperatures is caused mainly by the incipient destruction (denaturation) of the protein component of the enzyme. Since proteins in a dry state are denatured more slowly than hydrated proteins, which are usually in the form of a gel or solution, the inactivation of an enzyme occurs much more slowly in the dry state than in the presence of moisture. For this reason, dry bacteria spores or dry seeds are able to withstand far higher temperatures than wet spores or seeds.

The Danish physical chemist S. Sarensen was the first to establish that pH is an important influence on enzyme action. Enzymes vary according to the pH value that is optimal for their action. Pepsin, for example, which is contained in gastric juice, is most active in a strongly acidic medium (pH 1–2), and trypsin, a proteolytic enzyme secreted in the pancreas, has an optimum action in a weakly alkaline medium (pH 8–9). The optimum action of papain, a proteolytic enzyme of vegetable origin, occurs in a weakly acidic medium (pH 5–6).

Enzyme action also depends on specific activators and on nonspecific or specific inhibitors. For example, enterokinase, which is secreted in the pancreas, converts inactive trypsinogen into active trypsin. Similar inactive enzymes in the cells and secretions of various glands are called proenzymes. Many enzymes are activated by compounds containing the sulfhydryl group (—SH); among these compounds are the amino acid cysteine and the tri-peptide glutathione, which is found in every living cell. Glutathione has an especially strong activating effect on some proteolytic and oxidizing enzymes.

Nonspecific inhibition of enzymes is caused by substances that form insoluble precipitates with proteins or that block certain groups within proteins, for example, the—SH group. There are also specific inhibitors of enzymes; these inhibitors suppress catalytic functions by bonding with chemical groups in the enzyme's active centre. For example, carbon monoxide (CO) specifically inhibits a number of oxidizing enzymes that contain iron or copper in their active centres. Carbon monoxide forms a chemical compound with these metals, blocks the enzyme's active centre, and consequently halts the enzyme's activity.

The inhibition of enzymes can be reversible or irreversible. In reversible inhibition, for example, in the action of malonic acid on succinic dehydrogenase, the enzyme's activity is restored when the inhibitor is eliminated by dialysis or by another method. In irreversible inhibition the action of the inhibitor, even when the inhibitor is present in very low concentrations, is gradually intensified and eventually halts an enzyme's activity completely. The inhibition of enzyme action may also be competitive or noncompetitive. In competitive inhibition the inhibitor and the substrate seek to displace each other from the enzyme-substrate complex. High concentrations of a substrate reduce the action of a competitive inhibitor but maintain the action of a noncompetitive inhibitor. The action of activators and inhibitors on enzymes is of great importance in the regulation of enzymatic processes.

Classification and Nomenclature

In 1961 the International Union of Biochemistry recommended that enzymes be divided into six main classes: oxidoreductases, transferases, hydrolases, lyases, isomerases, and ligases. The following numbering system was also recommended: the number, or index, of each enzyme was to contain four digits, separated by periods. The first digit would indicate the class, the second the subclass, the third the sub-subclass, and the fourth the ordinal number in the sub-subclass. According to this system, the enzyme arginase, which decomposes arginine into ornithine and urea, has the number 3.5.3.1. In other words, arginase belongs to the hydrolase class, to the subclass of enzymes that act on the nonpeptide C—N bond, and to the sub-subclass of enzymes that decompose these bonds in linear compounds, rather than in cyclic compounds.

The oxidoreductase class includes enzymes that catalyze oxidation-reduction reactions. This class is divided into 14 subclasses according to the nature of the chemical group in the substrate molecule that undergoes oxidation, for example, the alcohol, aldehyde, or ketone group. The sub-subclasses of the oxidoreductases are numbered according to the type of hydrogen (electron) acceptor taking part in the reaction, for example, a coenzyme, cytochrome, or molecular-oxygen acceptor. The first three digits of the enzyme's number identify the type of enzyme; for example, 1.2.3 designates an oxidoreductase that acts as an electron acceptor on an aldehyde that contains molecular oxygen.

The transferase class, which includes enzymes that catalyze transfer reactions, is divided into eight subclasses according to the nature of the chemical groups transferred. These groups may be monocarbonic or glycosyl radicals, or they may be nitrogen or sulfur-containing groups. In the transferases, the third digit indicates the type of groups transferred; for example, the monocarbonic group may be methyl, carboxyl, or formyl.

The hydrolases are enzymes that catalyze the hydrolytic decomposition of various compounds. The hydrolases are divided into nine subclasses according to the type of bond hydrolyzed: ester, peptide, or glycoside. The third digit of a hydrolase's number defines the type of hydrolyzed bond.

The lyases are enzymes that detach a chemical group from the substrate nonhydrolytically, thus forming double bonds. Lyases also add chemical groups to double bonds. There are five subclasses of lyases. The second digit of a lyase's number refers to the type of bond that undergoes cleavage, for example, carbon-carbon or carbon-oxygen. The third digit refers to the type of chemical group detached.

Isomerases catalyze isomerization reactions and are divided into five subclasses according to the type of reaction catalyzed. Ligases, or synthetases, catalyze the joining together of two molecules, a coupling accompanied by the breakdown of a pyro-phosphate bond in a molecule of adenosine triphosphate (ATP) or of a similar triphosphate. The first digit of a ligase's number refers to the type of bond formed, for example, carbon-nitrogen or carbon-oxygen, and the second digit identifies the nature of the compound formed. Isomerases catalyze isomerization reactions and are divided into five subclasses according to the type of reaction catalyzed. Ligases, or synthetases, catalyze the joining together of two molecules, a coupling accompanied by the breakdown of a pyro-phosphate bond in a molecule of adenosine triphosphate (ATP) or of a similar triphosphate. The first digit of a ligase's number refers to the type of bond formed, for example, carbon-nitrogen or carbon-oxygen, and the second digit identifies the nature of the compound formed. Enzymes are also classified according to systematic and trivial (unsystematic) names. For example, the systematic name 2-oxo-acid carboxy-lyase corresponds to the trivial name pyruvate decarboxylase, and the systematic name L-arginine (amidinohydrolase), to the trivial name arginase.

Regulation of Enzymatic Activity

Living organisms have control mechanisms that regulate the synthesis and activity of enzymes. The complex of enzymes within a given organism is determined by the organism's genetic nature. However, this complex may be altered by such internal and external factors as diet, mutations, the effect of ionizing radiation, and the composition of the surrounding air. For example, such molecular diseases as alkaptonuria are caused by mutations. In this hereditary disease, changes in the amino acid tyrosine result in the formation of homogentisic acid. The homogentisic acid accumulates in the body and is excreted with the urine since patients with alkaptonuria lack the ability to synthesize the two enzymes that catalyze the subsequent oxidation of homogentisic acid: para-hydroxyphenyl-pyruvate oxidase and homogentisic acid oxidase.

The influence of nutriments on the organism's enzyme system is particularly easy to trace in microorganisms. For example, a colon bacillus (Escherichia coli) synthesizes only traces of P-galactosidase during its growth on a nutritive medium containing glucose. However, when various β-galactosides are present, large amounts of this enzyme are formed, amounting to 6-7 per cent of all the proteins contained in a cell. Enzymes whose formation or intensified synthesis is caused by the influence of a compound are called induced enzymes.

Other compounds may repress the synthesis of enzymes. In animals, the induction and repression of enzyme synthesis is influenced by hormones as well as by substrates and metabolites. For example, the synthesis of the enzyme glucose-6-phosphatase, which helps synthesize glucose in the liver, is induced by the hormones thyroxine and cortisone but is repressed by insulin. A general theory of the induction and repression of biosynthesis on the genetic level was formulated by the French biologists F. Jacob and J. Monod.

An enzyme may be present in an organism in different molecular forms. The distinct forms of an enzyme that catalyze the same reaction but that differ in physical, chemical, and immunological properties are called isoenzymes. The synthesis of isoenzymes is determined by genetic factors, but it may be altered by environmental factors. Thus, the concentration and activity of an enzyme depend on many diverse environmental factors. These include the water, air, temperature, acidity, and light, the concentration of substrates and of cofactors necessary for enzyme action, the presence of activators and inhibitors, the concentration of metabolites, and, in higher multicellular organisms, the neural and hormonal regulation of enzyme activity.

The Pasteur effect, or the inhibition of fermentation by the action of oxygen, is an example of the effect of the environment on enzyme activity. The activity of many enzymes is regulated according to the allosteric principle. Such enzymes have an allosteric site that is joined to a specific metabolite. The allosteric effector causes changes in the structure of the active centre that reduce or increase the activity of the enzyme.

Some enzymes exist in a cell as components of complexes containing a number of enzymes. In such complexes the activity of each enzyme is controlled by the other enzymes in the complex. An example of such a complex is pyruvate dehydrogenase, which is composed of 16 molecules of pyruvate decarboxylase, eight molecules of dihydrolipoil dehydrogenase, and four aggregates of lipoate acetyltransferase, each of which consists of 16 subunits. Such subcellular structures as the mitochondria, the microsomes, and the lysosomes, as well as the proteolipide membranes that separate these structures from the cytoplasm, help regulate enzyme activity in the cell. Many enzymes are embedded in these membranes as part of complexes of enzymes.

Practical Importance of Enzymes

The fermentation process is fundamental to many industries, including the bread-baking, wine-making, beer-brewing, cheese-making, and distilling industries and is also fundamental to the production of tea and vinegar. Beginning in the early 20th century, the distilling industry was one of a number of industries to use the fermentation methods developed by the Japanese chemist J. Takamine. These methods made use of enzymatic agents obtained from molds or bacteria. Fermentation methods are widely used in a number of countries to saccharify starch by means of amylase into crystalline glucose or to convert starch into alcohol. When added to bread dough, concentrated amylolytic solutions of enzymes derived from molds improve the quality of the bread and reduce the time needed for rising. Solutions containing proteolytic enzymes derived from microorganisms are used in the leather industry to remove hair and soften hides, and to replace the expensive proteolytic enzyme rennin in the cheese-making industry. Solutions containing microbial pectolytic enzymes are widely used in the preparation of fruit juices; they increase the quantity of juice produced by 10-20 per cent.

Purified enzyme preparations are increasingly used in medicine. In both scientific research and clinical practice, highly purified enzyme preparations are used in biochemical analysis. The use of immobilized enzymes, that is, insoluble complexes of enzymes with some carriers, will undoubtedly increase.

The selection of an appropriate carrier results in the formation of a highly active immobilized enzyme that is resistant to denaturants. A column filled with the immobilized enzyme may be used repeatedly to produce a corresponding reaction. Immobilized enzymes are increasingly used in biochemical analysis and in biochemical technology

BIOREACTOR OPERATION PARAMETRES

Oxygen transfer, pH, and temperature, which are the major bioreactor operation parametres, show diverse effects on product formation in aerobic fermentation processes by influencing metabolic pathways and changing metabolic fluxes.

TEMPERATURE

Generally, for a given type of bacterium, growth proceeds most rapidly at a particular temperature: the optimum growth temperature; the rate of growth tails off as temperatures increase or decrease from the optimum. Rates of all reactions, including those catalyzed by enzymes, rise with increase in temperature in accordance with the Arrhenius equation:

$$k = Ae^{-\Delta G/RT}$$

Typical standard free energies of activation give rise to increases in rate by factors between 1.2 and 2.5 for every 10°C rise in temperature. So, it would be preferable to use enzymes at high temperatures in order to make use of this increased rate of reaction plus the protection it affords against microbial contamination. For enzymes, however, denaturation begins to occur at 45 to 50°C and is severe at 55°C. One physical mechanism for this phenomenon is obvious: as the temperature increases, the atoms in the enzyme molecule have greater energies and a greater tendency to move. Eventually, they acquire sufficient energy to overcome the weak interactions holding the globular protein structure together, and deactivation follows.

Temperature also affects product formation. However, the optimum temperature for growth and product formation may be different. When temperature is increased above the optimum temperature, the maintenance requirements of cells increase. The yield coefficient is also affected by temperature. Temperature also may affect the rate-limiting step in a fermentation process. At high temperatures, the rate of bioreaction might become higher than the diffusion rate, and diffusion would then become the rate-limiting step. In literature, the production of *E. coli* is performed at 37°C, which is the bacteria's optimal temperature of growth.

Ph

Enzymes are amphoteric molecules containing a large number of acidic and basic groups, situated mainly on the surface. The charges on these groups will vary, according to their acid dissociation constants, with the pH of their environment. This will affect the total net charge of the enzyme and the distribution of charge on its exterior surface, in addition to the reactivity of the catalytically active groups. Thus, the catalytically active enzyme may be a large or small fraction of the total enzyme present, depending upon the pH.

In a similar manner, the charge and the charge distribution on the substrate(s), product(s) and coenzymes (where applicable) will also be affected by pH changes. These charge variations, plus any consequent structural alterations, may be reflected in changes in the binding of the substrate, the catalytic efficiency and the amount of active enzyme. For these reasons, enzymes have a characteristic pH at which their activity is maximal; above or below this pH the activity declines. However, the pHactivity profiles of enzymes are not always bell-shaped. The optimum pH of an enzyme is not necessarily identical with the pH of its normal intracellular surroundings, which may be on the ascending or descending slope of its pHactivity profile. This fact suggests that the pH-activity relationship of an enzyme may be a factor in intracellular control of enzymatic activity. For many bacteria, pH optima ranges from 3 to 8; and many organisms have mechanisms to maintain intracellular pH at a relatively constant level in the presence of fluctuations in environmental pH. When pH differs from the optimal value, the maintenance-energy requirements increase.

In most fermentations, pH can vary substantially. Often the nature of the nitrogen source can be important. Furthermore, pH can change due to the production of organic acids, utilization of acids (particularly amino acids), or the production of bases. Thus, pH control by means of a buffer or an active pH control system is important. Nevertheless, some bioprocesses require controlled pH conditions, while others might require uncontrolled pH operations, in order to increase the product yield and selectivity. The studies in the literature using *E. coli* as a host microorganism, differs in the pH value they utilize. The effect of pH on the growth of recombinant *E. coli* at three different controlled pH values, *i.e.*, 7.0, 7.4, 8.0; and reported that cell concentration increases when the culture pH decreases, however, enzyme activity attain its highest value at pH values in the vicinity of pH=7.4. The pH constant at 7.0 with aqueous NH_3. pH was kept constant throughout the bioprocess at 7.4 with 10 N NaOH. Also, in some studies of Akesson *et al.* (pH=6.9 by 25% ammonia) and (pH=7.0 by 2 M NaOH) pH value was kept constant throughout the bioprocesses. Hence, there is no systematic investigation in the literature, on the effects of pH on any biomolecule production and by-product distributions by the host *E. coli*.

Oxygen Transfer

The Importance and Mechanism of Oxygen Transfer

As microorganisms require oxygen for their metabolic activities like respiration, growth and product formation in aerobic fermentations, the concentration of the dissolved oxygen in the medium and oxygen uptake rate are very important parametres. Oxygen is present in all organic cell components and cellular water and constitutes about 20% of the dry weight of the cells. Molecular oxygen is required as a terminal electron acceptor in the aerobic metabolism.

Gaseous oxygen is introduced into growth media by sparging air; and oxygen transfer rate can be adjusted by either changing the air inlet rate or agitation rate. Cells in aerobic culture take up oxygen from the liquid. The rate of oxygen transfer from gas to liquid is therefore of prime importance, especially at high cell densities when cell growth is likely to be limited by availability of oxygen in the medium. According to the two-film theory, the oxygen must pass through a series of transport resistances, the relative magnitudes of which depend on bubble hydrodynamics, temperature, cellular activity and density, solution composition, interfacial phenomena, and other factors.

- Transfer from the interior of the bubble to the gas-liquid interface,
- Movement across the gas-liquid interface,
- Diffusion through the relatively stagnant liquid film surrounding the bubble,
- Transport through the bulk liquid,
- Diffusion through the relatively stagnant liquid film surrounding the cells,
- Movement across the liquid-cell interface,
- If the cells are in a floc, clump or solid particle, diffusion through the solid to the individual cell,
- Transport through the cytoplasm to the site of reaction.

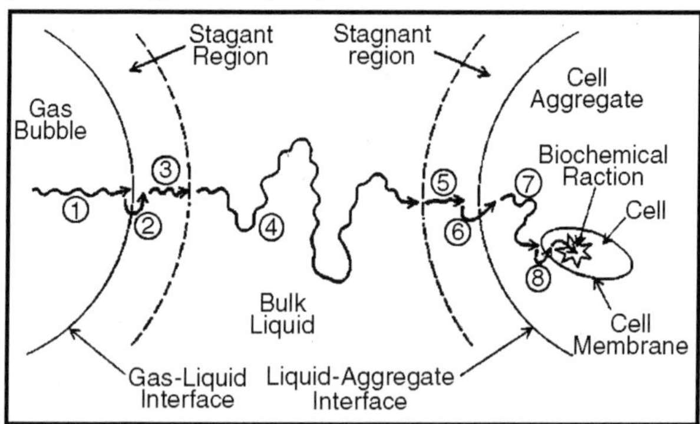

Fig. Schematic Diagram of Steps Involved in Transport of Oxygen from a Gas Bubble to Inside a cell.

When cells are dispersed in the liquid, and the bulk fermentation broth is well mixed, the major resistance to oxygen transfer is the liquid film surrounding the gas bubbles; therefore the rate of oxygen transfer from gas to liquid is of prime importance. An expression for oxygen transfer rate (OTR) from gas to liquid is given by the following equation:

$$OTR = k_L a\left(C_o^* - C_o\right)$$

where, k_L is the oxygen transfer coefficient, a is the gas-liquid interfacial area, kLa is the volumetric oxygen transfer coefficient, C_O* is saturated dissolved oxygen concentration, C_O is the actual dissolved oxygen concentration in the broth.

Since solubility of oxygen in aqueous solutions is very low, the liquid phase mass transfer resistance dominates, and the overall liquid phase mass transfer coefficient, K_La, is approximately equal to liquid phase mass transfer coefficient, k_La.

Oxygen Transfer Characteristics

The transfer of oxygen into the microbial cell in aerobic fermentation processes strongly affects product formation by influencing metabolic pathways and changing metabolic fluxes; therefore, to fine-tune bioreactor performance in relation with the physiology of the microorganism the extent of oxygen transfer requirement needs to be clarified. The rate of oxygen transfer in fermentation broths is influenced by several physical and chemical factors that change either the value of K_La, or the driving force for mass transfer, (C_O *– C_O). Therefore, the oxygen uptake rate of the cells, and the liquid phase mass transfer coefficient, K_La, are important characteristics of oxygen transfer.

The rate at which oxygen is consumed by cells in fermenters determines the rate at which it must be transferred from gas to liquid. Many factors influence oxygen demand; the most important of which are cell species, culture growth phase, and nature of the carbon source in the medium. In batch culture, rate of oxygen uptake varies with time. The reasons for this are twofold. First, the concentration of cells increases during the course of batch culture and the total rate of oxygen uptake is proportional to the number of cells present. In addition, the rate of oxygen consumption per cell, known as the specific oxygen uptake rate (q_O), also varies. Oxygen uptake rate (OUR), $-r_O$, per unit volume of broth is given by:

$$-r_o = q_o C_x$$

The inherent demand of an organism for oxygen (q_O) depends primarily on the biochemical nature of the cell and its nutritional environment. Choice of substrate for fermentation can also significantly affect oxygen demand. Because glucose is generally consumed more rapidly than other sugars or carboncontaining substrates, rates of oxygen demand are higher when glucose is used. The liquid phase mass transfer coefficient, KLa, is an important parametre in bioreactors, which indicates the oxygen transfer rate from gas to the liquid phase. It is important to estimate whether the dissolved oxygen in the medium is successfully transferred to the cells, and it is possible by determining the liquid phase mass transfer coefficient, KLa. Dynamic method is widely used for the determination of the value of KLa experimentally, and it can be applied during the fermentation process.

INDUSTRIAL USE OF AN IMMOBILIZED ENZYME

APPLICATIONS OF IMMOBILIZED ENZYMES

The first industrial use of an immobilized enzyme is amino acid acylase by Tanabe Seiyaku Company, Japan, for the resolution of recemic mixtures of chemically synthesized amino acids. Amino acid acylase catalyses the deacetylation of the L form of the *N*-acetyl amino acids leaving unaltered the *N*-acetyl-d amino acid, that can be easily separated, racemized and recycled. Some of the immobilized preparations used for this purpose include enzyme immobilized by ionic binding to DEAE-sephadex and the enzyme entrapped as microdroplets of its aqueous solution into fibres of cellulose triacetate by means of fibre wet spinning developed by Snam Progetti. Rohm GmbH have immobilized this enzyme on macroporous beads made of flexiglass-like material. By far, the most important application of immobilized enzymes in industry is for the conversion of glucose syrups to high fructose syrups by the enzyme glucose isomerase. It is evident that most of the commercial preparations use either the adsorption or the cross-linking technique. Application of glucose isomerase technology has gained considerable importance, especially in nontropical countries that have abundant starch raw material. Unlike these countries, in tropical countries like India, where sugarcane cultivation is abundant, the high fructose syrups can be obtained by a simpler process of hydrolysis of sucrose using invertase. Compared to sucrose, invert sugar has a higher humectancy, higher solubility and osmotic pressure.

Historically, invertase is perhaps the first reported enzyme in an immobilized form. A large number of immobilized invertase systems have been patented. The possible use of whole cells of yeast as a source of invertase was demonstrated by D'Souza and Nadkarni as early as 1978. A systematic study has been carried out in our laboratory for the preparation of invert sugar using immobilized invertase or the whole cells of yeast. These comprehensive studies carried out on various aspects in our laboratory of utilizing immobilized whole-yeast have resulted in an industrial process for the production of invert sugar.

Table. Commercial Immobilized Lactase Preparetions

Company procedure	Enzyme source	Immobilization
Gist brocades	Saceharomyces lactis triacetate fibres	Enzyme cntrapped in cellulose

Snampogetti	Klayveromyces luctis triacetate fibres	Enzyme entrapped in cellulose
Corning glass works	Aspergillus niger	Enzyme covalently bound to silica beads
Valio laboratory	A. niger phenol formaldehyde resin	Enzyme absorbed/cross-linked to
Sturge	A. niger to silanized Mn-Zn ferrite	Enzyme convalently bound
Rchm GmbH	A. oryzac macroporous plexiglass beads	Enzyme convalently bound to
Sumitomo	A. oryzac	Enzyme covalently bound to macroporous amphoteric ion exchange resin
Amerace Corp.	A. oryzac	Enzyme covalently bound to microporous PVC-silica sheet

L-aspartic acid is widely used in medicines and as a food additive. The enzyme aspartase catalyses a one-step stereospecific addition of ammonia to the double bond of fumaric acid. The enzymes have been immobilized using the whole cells of *Escherichia coli.* This is considered as the first industrial application of an immobilized microbial cell. The initial process made use of polyacrylamide entrapment which was later substituted with the carragenan treated with glutaraldehyde and hexamethylenediamine. Kyowa Hakko Kogyo Co. uses Duolite A7, a phenolformaldehyde resin, for adsorbing aspartase used in their continuous process.

Other firms include Mitsubishi Petrochemical Co. and Purification Engineering Inc. Some of the firms, specially in Japan like Tanabe Seiyaku and Kyowa Hakko, have used the immobilized fumarase for the production of malic acid (for pharmaceutical use). These processes make use of immobilized nonviable cells of *Brevibacterium ammoniagenes* or *B. flavus* as a source of fumarase. Malic acid is becoming of greater market interest as food acidulant in competition with citric acid. Studies from our laboratory have shown the possibility of using immobilized mitochondria as a source of fumarase.

One of the major applications of immobilized biocatalysts in dairy industry is in the preparation of lactose-hydrolysed milk and whey, using b - galactosidase. A large population of lactose intolerants can consume lactose-hydrolysed milk. This is of great significance in a country like India where lactose intolerance is quite prevalent.

Lactose hydrolysis also enhances the sweetness and solubility of the sugars, and can find future potentials in preparation of a variety of dairy products. Lactose-hydrolysed whey may be used as a component of whey-based beverages, leavening agents, feed stuffs, or may be fermented to produce ethanol and yeast, thus converting an inexpensive byproduct into a highly nutritious, good quality food ingredient. The first company to commercially hydrolyse lactose in milk by immobilized lactase was Centrale del Latte of Milan, Italy, utilizing the Snamprogetti technology. The process makes use of a neutral lactase from yeast entrapped in synthetic fibres.

Specialist Dairy Ingredients, a joint venture between the Milk Marketing Board of England and Wales and Corning, had set up an immobilized b -galctosidase plant in North Wales for the production of lactose-hydrolysed whey. Unlike the milk, the acidic b -galactosidase of fungal origin has been used for this purpose. An immobilized preparation obtained by cross-linking b -galactosidase in hen egg white (lyophilized dry powder) has been used in our laboratory for the hydrolysis of lactose. A major problem in the large-scale continuous processing of milk using immobilized enzyme is the microbial contamination which has necessitated the introduction of intermittent sanitation steps.

A co-immobilizate obtained by binding of glucose oxidase on the microbial cell wall using Con A has been used to minimize the bacterial contamination during the continuous hydrolysis of lactose by the initiation of the natural lactoperoxidase system in milk. A novel technique for the removal of lactose by heterogeneous fermentation of the milk using immobilized viable cells of *K. fragilis* has also been developed.

One of the major applications of immobilized enzymes in pharmaceutical industry is the production of 6-aminopenicillanic acid (6-APA) by the deacylation of the side chain in either penicillin G or V, using penicillin acylase (penicillin amidase). More than 50% of 6-APA produced today is enzymatically using the immobilized route. One of the major reasons for its success is in obtaining a purer product, thereby minimizing the purification costs. The first setting up of industrial process for the production of 6-APA was in 1970s simultaneously by Squibb (USA), Astra (Sweden) and Riga Biochemical Plant (USSR). Currently, most of the pharmaceutical giants make use of this technology. A number of immobilized systems have been patented or commercially produced for penicillin acylase which make use of a variety of techniques either using the isolated enzyme or the whole cells. This is also one of the major applications of the immobilized enzyme technology in India. Similar approach has also been used for the production of 7-aminodeacetoxy-cephalosporanic acid, an intermediate in the production of semisynthetic cephalosporins.

Immobilized oxidoreductases are gaining considerable importance in biotechnology to carry out synthetic transformations. Of particular significance in this regard are oxidoreductase-mediated asymmetric synthesis of amino acids, steroids and other pharmaceuticals and a host of speciality chemicals. They play a major role in clinical diagnosis and other analytical applications like the biosensors. Future applications for oxidoreductases can be in areas as diverse as polymer synthesis, pollution control, and oxygenation of hydrocarbons. Immobilized glucose oxidase can find application in the production of gluconic acid, removal of oxygen from beverages, and in the removal of glucose from eggs prior to dehydration in order to prevent Maillard reaction. Studies carried out in this direction in our laboratory have shown that glucose can be removed from egg, using glucose oxidase and catalase which are co-immobilized either on polycationic cotton cloth or in hen egg white foam matrix.

Alternatively, glucose can also be removed by rapid heterogeneous fermentation of egg melange, using immobilized yeast. Immobilized D-amino acid oxidase has been investigated for the production of keto acid analogues of the amino acids, which find application in the management of chronic uremia. Keto acids can be obtained using either L- or D-amino acid oxidases. The use of D-amino acid oxidase has the advantage of simultaneous separation of natural L-isomer from DL-recemates along with the conversion of D-isomer to the corresponding keto acid which can then be transaminated in the body to give the Lamino acid.

Of the several microorganisms screened, the triangular yeast *T. variabilis* was found to be the most potent source of D-amino acid oxidase with the ability to deaminate most of the D-amino acids. The permeabilized cells entrapped either in radiation polymerized acrylamide Ca-alginate or gelatin have shown promise in the preparation of a -keto acids. Another interesting enzyme that can be used profitably in immobilized form is catalase for the destruction of hydrogen peroxide employed in the cold sterilization of milk. A few reports are available on its immobilization using yeast cells.

Lipase catalyses a series of different reactions. Although they were designed by nature to cleave the ester bonds of triacylglycerols (hydrolysis), lipase are also able to catalyse the reverse reaction under microaqueous conditions, *viz.* formation of ester bonds between alcohol and carboxylic acid moieties. These two basic processes can be combined in a sequential fashion to give rise to a set of reactions generally termed as interesterification. Immobilized lipases have been investigated for both these processes.

Lipases possess a variety of industrial potentials starting from use in detergents; leather treatment controlled hydrolysis of milk fat for acceleration of cheese ripening; hydrolysis, glycerolysis and alcoholysis of bulk fats and oils; production of optically pure compounds, flavours, etc. Lipases are spontaneously soluble in aqueous phase but their natural substrates (lipids) are not. Although use of proper organic solvents as an emulsifier helps in overcoming the problem of intimate contact between the substrate and enzyme, the practical use of lipases in such psuedohomogeneous reactions poses technological difficulties. Varieties of approaches to solve these, using immobilized lipases, have recently been reviewed.

Significant research has also been carried out on the immobilization and use of glucoamylase. This is an example of an immobilized enzyme that probably is not competitive with the free enzyme and hence has not found large-scale industrial application. This is mainly because soluble enzyme is cheap and has been used for over two decades in a very optimized process without technical problems. Immobilization has also not found to significantly enhance the thermostability of amylase. Immobilized renin or other proteases might allow for the continuous coagulation of milk for cheese manufacture. One of the major limitations in the use of enzymes which act on macromolecular substrates or particulate or colloidal substrates like starch or cellulose pectin or proteins has been the low retention of their realistic activities with natural substrates due to the steric hindrance.

Efforts have been made to minimize these problems by attaching enzymes through spacer arms. In this direction, application of tris (hydroxymethyl) phosphine as a coupling agentmay have future potentials for the immobilization of enzymes which act on macromolecular substrates. Other problem, when particulate materials are used as the substrates for an enzyme, is difficulty in the separation of the immobilized enzyme from the final mixture. Efforts have been made in this direction to magnetize the bicatalyst either by directly binding the enzyme on magnetic materials (magnetite or stainless steel powder) or by co-entrapping magnetic material so that they can be recovered using an external magnet. Magnetized biocatalysts also help in the fabrication of magnetofluidized bed reactor.

A variety of biologically active peptides are gaining importance in various fields including in pharmaceuti-cal industries and in food industries as sweeteners, flavourings, antioxidants and nutritional supplements. Proteases have emerged over the last two decades as powerful catalysts for the synthesis and modification of peptides. The field of immobilized proteases may have a future role in this area. One of the important large scale applications will be in the synthesis of peptide sweetener using immobilized enzymes like the thermolysin. Proteolytic enzymes, such as subtilisin, *a*-chymotrypsin, papain, ficin or bromelain, which have been immobilized by covalent binding, adsorption or cross-linking to polymeric supports are used (Bayer AG) to resolve A *N*-acyl-DL-phenylglycine ester racemate, yielding *N*-acyl-D-esters or *N*-acyl-D-amides and *N*-acyl-L-acids. Immobilized aminopeptidases have been used to separate DL-phenylgycinamide racemates. SNAM-Progetti SpA-UK have used the immobilized hydropyrimidine hydrolase to prepare D-carmamyl amino acids and the corresponding D-amino acids from various substituted hydantoins.

4

Enzymes in Medicine

Since the mid-1950s there has been a considerable increase in both measurement of enzyme activities and the use of purified enzymes in clinical practice. In recent years, many enzymes have been isolated and purified, and this made it possible to use enzymes to determine the concentration of substrates and products of clinical importance. A further development, arising from the increased availability of purified enzymes, has been targeted to enzyme therapy.

CLINICAL DIAGNOSIS

The measurement of enzyme activities in serum is of major importance as an aid in diagnosis, being used as means of monitoring progress after therapy, recovery after surgery, and detection of transplant rejection. Urine can be also analysed for determination of enzyme activities since the detection of certain enzymes in urine may indicate kidney damage or failure. On the other hand, the concentration of certain metabolites in serum or urine may be determined by enzymatic methods. The method consists of using an enzyme to transform a metabolite into its product and then estimate the amount of transformed substrate. The use of enzymatic methods presents several advantages, such as the high specificity of the enzyme to estimate the concentration of the metabolite in the presence of other substances, avoiding the need of purification steps prior to chemical analysis. In addition, enzymatic reactions are performed at mild conditions, allowing the analysis of labile compounds that would be degraded by harsher chemical methods. The cost of purified enzymes may be, however, too high to support routine analysis, but the use of immobilized enzymes allows for enzyme reuse and the application of immobilized enzymes for diagnostic assays and as biosensors. The determination of serum metabolites in serum by enzymatic methods includes a wide range of substances such as glucose, uric acid, urea, cholesterol, cholesterol esters, triglycerides, and creatine, among others.

Enzyme Therapy

Many inborn metabolic disorders are associated with the absence of activity of one particular enzyme normally found in the body. Of the 1250 autosomal recessive human genetic diseases, over 200 involve errors in metabolism that result from specific known enzyme deficiencies. The initial identification of the disease may be difficult to determine and normally requires a tissue biopsy. For some genetic diseases, DNA probes are now available that can be used on small amounts of blood, cells, or amniotic fluid. Albinism, for example, is often caused by the absence of tyrosinase, an enzyme essential for the production of cellular pigments. Tyrosinase is a coppercontaining enzyme that catalyzes the first two rate-limiting steps in the melanin biosynthetic pathway, the oxidation of tyrosine to dopa and the subsequent dehydrogenation of dopa to dopaquinone.

The human tyrosinase gene, encoding 529 amino acids, consists of five exons spanning more than 50 Kb of DNA in chromosomes. When homozygous mutations of the tyrosine gene result in complete absence of melanogenic activity, such a patient, categorized as tyrosinasenegative oculocutaneous albinism, will never develop any melanin pigment in the skin, hair, and eyes throughout his or her life.

Some inborn errors in metabolism are relatively harmless, *e.g.*, albinism or alkaptonuria, but others must be detected early if the defect is to be circumvented. This is the case of phenylketonuria, where the enzyme phenylalanine 4-monooxygenase (enzyme that converts phenylalanine into tyrosine) is missing.

Phenylketonuria results in the accumulation of phenylalanine, which may cause mental retardation. Patients with phenylalanine 4-monooxygenase deficiency must follow a phenylalanine-free diet in order to avoid the accumulation of deleterious effects.

Phenylalanine is, however, an amino acid essential to maintain growth and protein turnover, and thus it must be supplied in a minimal amount required to maintain normal metabolism. Such a dietary scheme is normally carried out by lowering the amount of protein consumed, but this may cause a deficiency in other essential amino acids. A possible alternative to this therapy is to replace the missing enzyme. It may be difficult to find, however, an enzyme with the same function from a human source since the direct administration of enzymes from other sources into the body would cause an adverse immunological response. A possible approach to circumvent this problem involves the isolation of the enzyme within a microcapsule, fibre, or gel, which will protect the enzyme from proteolysis and avoid the undesirable immunological response.

The pharmacological properties of enzymes have been employed to replace enzymes that are missing or defective as a consequence of an inherited disease or malfunction of an organ where they are normally synthesized or to accomplish a certain biological effect that is dependent on the catalytic activity of the enzyme.

Therefore, depending on the treatment, the administration of enzymes as therapeutic agents can be subdivided into two categories:

- The topical application of an enzyme as an extracellular agent
- The intracellular applications of enzymes to treat metabolic deficiency and related disease.

The main areas where enzyme therapy has been applied are the degradation of necrotic tissue by the use of proteolytic enzymes, removal of toxic compounds from the blood, treatment of genetic deficiency diseases and cancer, and treatment of pancreatic insufficiency. Enzymes may be administered either intra- or extracorporeally, depending on the objective. If the enzyme is to be used for the removal or transformation of a substance present in the blood (*e.g.*, toxic metabolite or a blood clot), then it is only necessary for the enzyme to be present in the blood and not necessary for the enzyme to enter the intracellular compartments.

This type of application may be either intra- or extracorporeal using a bypass as in kidney dialysis. For intracellular therapy, it is necessary for the enzyme to be taken up by the appropriate target cells. Although the attempts made with enzyme therapy in clinical trials have so far had limited success, it is reasonable to assume that the delivery of enzymes would constitute a feasible approach for the treatment of certain diseases in the near future.

BIOREACTORS FOR EXTRACORPOREAL ENZYME THERAPY

Extracorporeal shunts have been proposed for the treatment of several clinical conditions. The most likely applications for enzymatic treatment are the removal of urea during kidney failure, removal of toxins (*e.g.*, paracetamol) during liver failure, or the reduction of key metabolites from the circulation to treat cancer. Urease is one of the most important enzymes in biomedical applications. Urease is an enzyme that catalyzes the hydrolysis of urea to form ammonia and carbon dioxide. Urea is one of the main metabolic end products, and the removal of its excess has been a major problem for patients suffering from renal failure. Hence, its immobilization by entrapment has been investigated by many workers for applications in biosensors and as artificial kidneys.

The most attention has been given to the development of enzyme reactors, where the urea would be removed and the dialysis fluid prepared for further use. The use of this enzyme is often limited due to its high cost, availability in small amounts, instability, and the limited possibility of feasible recovery of these biocatalysts from a reaction mixture. Numerous synthetic and natural polymeric supports have been used for urease immobilization, and their uses in medical and technical fields are well reported. The covalent bond of urease in different supports has been reported in many studies. Some commonly used supports are chitosan-poly(glycidil methacrylate), carboxy-methylcellulose, polyurethane, sepharose-2B, polyacrylamide, ion exchange resins, copolymers of polyglycidylmethacrylate, calcium alginate beads, poly(vinyl alcohol) (PVA), hydroxyapatite, 2-dimethylaminoethylmethacrylate, poly(ethylene glycol dimethacry-late/2-hydroxy ethylene methacrylate) microbeads, poly(caprolactone)/starch, and poly(orthoesters).

L-asparaginase has been used for treating leukemias and disseminating cancers that require asparagines for growth, but this treatment presents several serious side effects. To overcome these problems, immobilized enzyme derivatives have been prepared on various supports for extracorporeal treatment. Blood can be passed over the immobilized enzyme, thus depleting the asparagine supply needed by the cancer cells. The enzyme does not come into direct contact with the organs to which it is toxic, and hypersensitivity reactions do not occur. With this type of treatment, however, the blood plasma must be first separated from the cells to minimize cell damage and then passed through a separate column containing the immobilized enzyme.

This process requires that the blood remain outside the body for relatively long periods of time, resulting in the denaturation and depletion of many plasma proteins. Some techniques have been developed to minimize this problem, such as the use of a porous hollow-fibre plasmapheresis device. With this system, the plasma can be separated from the whole blood and contact with immobilized enzyme in one passage, thus minimizing its time outside the body and then reducing the damage to the plasma proteins. Adsorption techniques have been used to immobilize asparaginase onto hollow fibres after first coating the fibres with albumin and then crosslinking the enzyme with glutaraldehyde.

Maciel and Minim also reported that the use of *L*-asparaginase covalently attached to nylon tubing may constitute a useful system to be used in clinical applications. Bilirubin oxidase is also an example of enzyme used in extracorporeal applications. All human newborns accumulate bilirubin to levels greater that those in adults, and 20% accumulate enough to stain their skin, resulting in jaundice. Bilirubin binds to cellular and mitochondrial membranes, causing cell death in a variety of tissues. Clinically, bilirubin toxicity may lead to mental retardation, cerebral palsy, deafness, seizures, or death. The most common treatments for jaundiced infants are phototherapy and exchange transfusion.

This technique presents serious problems such as hypoglycemia, hypocalcemia, acidosis, transmission of infectious, etc. Lavin *et al.* reported the use of a highly specific enzyme to remove bilirubin from the bloodstream using a small reactor (extracorporeal circuit) containing bilirubin oxidase covalently immobilized in agarose beads. These researchers obtained good results for the removal of the bilirubin in humans and in genetically jaundiced rats.

Heparinase, an enzyme that degrades heparin into small polysaccharides, has also been immobilized into an extracorporeal device (artificial kidney bioreactor) to eliminate the anticoagulant properties of heparin (used to prevent clotting in the device) before the blood returns to the patient.

Diagnostic Assays and Biosensors

Isolated or combined enzymes are being used in medicine as useful tools for clinical analysis. About 50 different enzymes are used in different aspects of clinical diagnoses, and for most of these, much higher levels of purity are required than for most industrial enzymes. Two of the major enzymes used are peroxidase from horseradish and alkaline phosphatase from beef intestinal mucosa, both being required for immunoassays. The enzymes may be used in test strips, ELISA, biosensors, and autoanalysers. The serum uric acid concentration is an important index for clinical diagnosis of gout, leukemia, toxemia of pregnancy, and severe renal impairment.

A number of enzymes are assayed in serum and urine for diagnostic purposes; the more frequently used ones are discussed below. Alkaline phosphatases (ALP) are a group of enzymes found primarily in the liver (isoenzyme ALP-1) and bone (isoenzyme ALP-2). There are also small amounts produced by cells lining the intestines (isoenzyme ALP-3), the placenta, and the kidney (in the proximal convoluted tubules). What is measured in the blood is the total amount of alkaline phosphatase released from these tissues into the blood. As the name implies, this enzyme works best at an alkaline pH (pH 10), and thus the enzyme itself is inactive in the blood. Alkaline phosphatase acts by splitting off phosphorus (an acidic mineral), creating an alkaline pH.

The primary importance of measuring alkaline phosphatase is to check the possibility of bone or liver diseases. Another application of immobilized enzymes is the development of improved sensing devices. Because of their high specificity for given substances, enzymes and monoclonal antibodies are particularly suitable for use as sensors. The membrane-covered electrode described by Clark in 1959 is the dominating sensor for the measurement of dissolved oxygen. Numerous modifications of the original concept have been developed. For instance, biosensors using enzymes have been used to detect the presence of various organic compounds, and recent developments have proven to be both rapid and highly selective. They have been used in important applications such as in clinical laboratories, fermentation processes, and pollution monitoring. Most of them have used a free or immobilized enzyme and an ion-sensitive electrode that measures indirectly (*e.g.*, by temperature or colour changes produced by an enzymatic reaction) the presence of a product whose formation is catalyzed by the enzyme.

The biosensors usually have immobilized biological molecules attached to the surface of a transducer that allows an electronic or optical signal to be converted into an appropriate signal. This type of biosensor could be used to measure glucose, sucrose, lactose, *L*-lactate, galactose, *L*-glutamate, *L*-

glutamine, choline, ethanol, methanol, hydrogen peroxide, starch, uric acid, etc., by using specific enzymes. Glucose oxidase is normally used to assay glucose concentration. Glucose sensors are the biosensors that have attracted much interest in both research and applications fields. One particularly important medical application of improved biosensors could be in the treatment of diabetic patients for whom proper levels of insulin and glucose must be maintained. For instance, small implantable devices for sampling blood to determine the levels of glucose and regulate the delivery of insulin could be developed using this enzyme.

A great variety of immobilization methods (*e.g.*, encapsulation, entrapment) and transducers have been developed to construct the glucose sensors with better performance and practicability since the work of Clark and Lyons in 1962. Several materials have been used, such as polyethylene terephthalate (PET), polyacrylamide, *N*- isopolyacrylamide, sol-gel, poly(2-hydroxyethyl methacrylate), alginate, artificial resins, glass, etc. Immobilized alkaline phosphatase on a nanoporous nickel-titanium film for sensor applications. A number of other enzymes have been described with great potential for medical applications, including carboxypeptidases, collagenase, fibrinolysin, pepsin, streptokinase, subtilisin, thrombin, tissue plasminogen activator, amylase, galactosidase, glucoamylase, lactase (-galactosidase), pectinase, pancreatin, phospholipases, cholesterol esterase and other DNases, RNases, phosphatases, esterases, sulfatases, isomerases, glucose isomerase, superoxide dismutase, cholesterol esterease, creatine kinase, and penicillin acylase.

Advantages and Disadvantages of Immobilized Enzymes

The use of immobilized enzymes normally offers several advantages over free enzymes, such as increased stability, localization, and retention of the molecules at the material surface, which enables easier handling, repeated use, and decreased cost. Other important advantages of using therapeutic immobilized enzymes are the prolonged blood circulation lifetime without the loss of specific activity and the lower immunogenicity. This advantage is particularly important for delivering enzymes or other biomolecules and may constitute an alternative and suitable method for the enzyme replacement therapy.

However, some limitations have been attributed to the use of immobilized enzymes in biomedical applications, such as mass transfer resistances (substrate in and product out), adverse biological responses of enzyme support surfaces (*in vivo* or *ex vivo*), fouling by other biomolecules, greater potential for product inhibition, and sterilization difficulties.

Although the preparation of sterile immobilized enzyme systems may be complex, sterilization may be achieved by filtrating all the reagents and protein solutions through 0.2m filters and working under aseptic conditions.

A very important issue regarding the use of enzymes or other products derived from biological or biotechnological processes in medical applications is to ensure that these therapeutic products do not contain any pyrogenic material, toxins, or infectious agents able to cause harmful effects.

For that, it is necessary to perform a complete examination of the products to test their safety in terms of local tolerance, toxicity, carcinogenicity, and immunogenicity, among other pharmacological safety tests. Taking into account the diversity in the range of products and the uncertainty about the regulatory status of some of them, it is necessary to design safety evaluation programmes to provide useful information to the responsible of clinical trials and to ensure patient safety.

ENZYME IMMOBILIZATION IN BIOMEDICAL APPLICATIONS

Biomaterials can be combined with biomolecules, such as enzymes and growth factors, to yield biologically functional systems. There is a wide and diverse range of materials and methods available for enzyme immobilization on or within the biomaterial. Furthermore, the criteria for selecting the immobilization methods should also take into account that the immobilized enzyme should retain an acceptable level of activity over a certain period of time in terms of economic or clinical aspects.

The methods used for the administration of immobilized enzymes may be divided into two principal groups: immobilized enzymes that are intended for prolonged circulation and enzymes that must be necessarily present in different tissues and organs of the body. In the second case, the immobilized enzyme is intended for local deposition during the treatment of discrete lesions (*e.g.*, thrombi, tumors, atherosclerotic injuries) or of the individual organs.

BIOLOGICALLY FUNCTIONAL SURFACES

Biomaterials, especially when used in tissue engineering applications, must have the capacity to induce tissue regeneration/repair in order to achieve a more rapid recovery of the defect. At present, the existing scaffolds are not satisfactory in achieving rapid and full recovery of the defect. The attachment of cells to biomaterials, and their subsequent spreading, are mediated by extracellular matrix (ECM) glycoproteins such as fibronectin, collagen, etc. ECM glycoproteins contain short sequences with cell attachment properties, which interact with the integrin family of cell surface receptors. The peptide sequence Arg-Gly-Asp (RGD) has been identified as being capable of interacting with cell surface receptors.

Thus, the immobilization of biologically active molecules on the surface of biomaterials for presenting effectors to target cells or to induce a particular effect is of great interest, since the immobilization of active agents presents the advantage of providing a continuous and localized stimulus for cell proliferation. Unlike nonimmobilized active agents, which are often consumed by cells, immobilized biomolecules remain bound to a substrate that is not consumed by cells and thus remain available to stimulate growth of additional cells.

This is particularly useful in perfusion cultures in which growth medium is continuously added and removed to allow long-term cell proliferation. Biodegradable polymers have been used as scaffolding materials for various tissue engineering applications because they can provide the support on which cells and tissues can adhere, but they can also guide and regulate the proliferation and activities of the adhered cells. However, the intrinsic hydrophobic property of some of these polymers restricts their applications as cell colonizing materials.

Many methods have been used to modify the properties of polymer surface, such as plasma treatment-induced grafting polymerization, ozone oxidation, and immobilization of enzymes, and special biologically active agents have been used to introduce reactive groups onto polymeric surfaces. Various strategies have been developed to incorporate bioactive agents on the surface of biomaterials for controlling cell and tissue responses.

The immobilization process can be involved, enriching surfaces for enhancing the cellular adhesion. Biomolecules such as enzymes, antibodies, antigens, peptides, or drugs have been immobilized on or within polymeric systems. An example of an adhesive protein is fibronectin, which is able to promote cellular adhesion through binding to integrin receptors, and this interaction has also been shown to play a role in cell growth, differentiation, and overall regulation of cell function.

Hern and Hubbell showed that the incorporation of the adhesion peptide RGD into a nonadhesive hydrogel proved to be useful for tissue resurfacing. There are numerous other adhesion peptides for targeting particularly desirable cell types and to modulate biological responses. Urokinase has been widely used for the clinical treatment of thrombogenetic disease and hemorrhoidal disease.

Artificial organ materials, on which urokinase was immobilized for its fibrinolytic activity, have been developed for blood-compatible materials. Immobilized urokinase by encapsulation in poly(2-hydroxyethyl methacrylate) and Introduced urokinase on the surface of the polytetrafluoroethylene using plasma modification technique by covalent bond. Another example of immobilized urokinase application was reported by Kato and coworkers, who had used urokinase immobilized in a Teflon catheter for treatment of thrombosis. Most of the studies found in the literature, regarding the surface functionalization of biomaterials with biological molecules, include the incorporation of adhesion and differentiation factors. The same approach may be used to incorporate specific enzymes able to regulate a number of cell functions. For instance, it is known that mitogen activated protein (MAP) kinase, upon activation by dual phosphorylation at threonine and tyrosine residues, is able to activate downstream targets that have been implicated in controlling gene expression, cell differentiation, and proliferation. This enzyme may be immobilized on the surface of biomaterials to control cell response, but other possibilities using different enzymes remain unexplored.

Enzyme Delivery

Although purified enzymes are now available for some enzyme deficiency diseases there are many problems in delivering the enzyme to the required site under such conditions that it will remain stable and active for a reasonable time. Normally, quite large amounts of enzyme are necessary with high level of purity and in a nonimmunogenic form. In addition, many enzymes when administered are inactivated or degraded fairly rapidly. The delivery of therapeutic molecules requires, therefore, efficient strategies to have a precise control on their release profile according to specific locations.

It might be possible to control the release of such molecules by creating delivery systems sensitive to changes in pH, temperature, or salt concentration or to the feedback provided by cells. The concept of enzyme-activated drugs in therapy is scientifically, as well as clinically, attractive, as it allows the chemist and enzymologist full intellectual rein in designing interlinked systems. As therapeutic drugs, enzymes possess several attributes such as high specificity towards substrate, high solubility for preparing liquid formulations, and optimum activity under physiological conditions. The administration of enzymes, in cases of enzyme deficiency and inborn errors of metabolism and in the treatment of certain types of cancer, appears to offer a successful form of therapy.

Cancer therapy based on the delivery of enzymes to tumor sites has advanced in several directions since antibody-directed enzyme/prodrug therapy was first described. Nanospheres, nanocapsules, liposomes, micelles, and other nanoparticulates are frequently referred to as carriers for delivery of therapeutic and diagnostics agents. Asparagine is an essential amino acid for certain types of leukemias that lack asparagines synthetase activity. The activity of *L* -asparaginase is to degrade asparagine into aspartate and ammonia. Therefore, asparaginase has been of interest to biochemists and clinicians as a possible cancer therapeutic agent. Some success has been achieved in administering asparaginase in capsules made of nylon and polyurea to mice and rats.

Although asparaginase has been found to be effective in the treatment of some patients, it may have several serious side effects. Relatively high concentrations of the enzyme are needed for it to be clinically effective. These levels cause a wide range of toxic effects on several organs including the liver, pancreas, kidneys, and brain. The enzyme may also be recognized as foreign by the body and potentially severe immunogenic responses will be stimulated, resulting in hypersensitivity reactions. To overcome these problems, immobilized enzyme derivatives have been prepared on various supports for extracorporeal treatment. Enzymes may be also used in cancer therapy as prodrug activators. This therapy consists basically in using a drug that has been chemically modified so that it remains inactive until specifically activated by an enzyme at the target site. Another example of an anticancer enzyme is hyaluronidase. Hyaluronidase is a globular enzyme of endoglycosidase action, which can depolymerize hyaluronic acid in the organism, decreasing its viscosity and increasing tissue permeability.

Hyaluronidase has been utilized extensively as an adjunct in anticancer chemotherapy regimens, suggesting that hyaluronidase has intrinsic anticancer properties against tumor growth. *In vitro* studies in tissue culture with tumor spheroids and *in vivo* tests using animal models demonstrated the beneficial effect of hyaluronidase for the penetration of drugs into tumor tissue. Later, in a prospective clinical trial, hyaluronidase significantly improved the outcomes of patients with bladder carcinoma if the enzyme was administrated topically together with mitomycin.

Hyaluronidase is also used for local application (subcutaneous injections) during treatment of joint disease, in dermatology, and in ophthalmology. Hyaluronidase had been used in ophthalmology with the aim of formation of a thinner scar and to prevent necrosis after paravataes with zytostatics. Trypsin has also been used to remove dead tissue from wounds, burns, and ulcers to speed the growth of new tissue and skin grafts, as well as to inhibit the growth of some contaminant organisms. The inappropriate activation of trypsinogen within the pancreas leads to development of pancreatitis. Once trypsin is activated, it is capable of activating many other digestive pro-enzymes.

These activated pancreatic enzymes further enhance the auto-digestion of the pancreas. Many materials, such as nylon, polysulfone, glycidyl methacrylate, chitosan, cellulose, and cellulose derivatives, have been used for trypsin immobilization. Enzyme therapy has also been tested for pancreatic insufficiency and cystic fibrosis. Pancreatic insufficiency can be alleviated by administrating orally enteric-coated microspheres containing lipase, amylase, and proteases. A special polymer coating protects the enzymes at low pH, such as in the stomach, and then releases them in the intestine at physiological pH. Lysozyme is a good example of an enzyme that catalyzes chemical reactions in the cell.

Lysozyme acts to kill bacteria by cleaving the covalent bond between the alternating polysaccharides that compose peptidoglycan in bacterial cell walls. The human salivary defence proteins and lysozyme are known to exert a wide antimicrobial activity against a number of bacterial, viral, and fungal pathogens *in vitro.* Therefore, these proteins, alone or in combinations, have been incorporated as preservatives in foods and pharmaceuticals as well as in oral health care products to restore saliva's own antimicrobial capacity in patients with dry mouth. These antimicrobials used in oral health care products, such as dentifrices, mouth rinses, moisturizing gels, and chewing gums, have been purified from bovine colostrum.

Other studies had been reported with lysozyme bound to chitosan, silica gel by means of physical adsorption, crosslinking to a polystyrene divinylbenzene matrix by the formation of ionic bindings, and by covalent attachment to nonporous glass beads. Harada and Kataoka described lysozyme immobilized into poly(ethylene glycol)–poly(aspartic acid) micelle. Lysozyme was selected as a model protein to incorporate into the micelle because it has a high isoelectric point (pI = 11), is positively charged over a wide range of pH, and has practical usage in drug delivery application as a lytic enzyme.

Chen and Chen prepared immobilized lysozyme by carbodiimide method to form amide bonds with an enteric coating polymer (hydroxypropyl methylcellulose acetate succinate [AS-L]) as the carrier, which shows reversibly soluble-insoluble characteristics with pH changes.

The glucose-6-phosphate dehydrogenase enzyme catalyzes the oxidation of glucose-6-phosphate to 6-phosphogluconate while concomitantly reducing the oxidized form of nicotinamide adenine dinucleotide phosphate (NADP+) to nicotinamide adenine dinucleotide phosphate (NADPH). NADPH, a required cofactor in many biosynthetic reactions, maintains glutathione in its reduced form. Reduced glutathione acts as a scavenger for dangerous oxidative metabolites in the cell. With the help of the enzyme glutathione peroxidase, reduced glutathione also converts harmful hydrogen peroxide to water. Red blood cells rely heavily on glucose-6-phosphate dehydrogenase activity because it is the only source of NADPH that protects the cells against oxidative stresses. People deficient in glucose-6-phosphate dehydrogenase are not prescribed, therefore, with oxidative drugs because their red blood cells undergo rapid hemolysis under this stress.

In Greece, glucose-6-phosphate dehydrogenase deficiency is the main cause of severe neonatal jaundice. The deficiency of this enzyme affects all races; the highest prevalence is among persons of African, Asian, or Mediterranean descent. Study of immobilized glucose-6-phosphate dehydrogenase has been reported.

Kotorman *et al.* immobilized glucose-6-phosphate dehydrogenase from yeast on polyacrylamide beads possessing carboxylic functional groups activated by a watersoluble carbodiimide. They verified highest operational stability of immobilized glucose-6-phosphate dehydrogenase. In relation to the use of immobilized enzymes with therapeutic purposes, catalase is one of the most interesting because it is employed to accelerate healing as well as to correct hereditary deficiencies and, in combination with hydrogen peroxide, as an antiseptic against anaerobes. The catalase enzyme has the ability to decompose hydrogen peroxide into oxygen and water, playing a central role in controlling the hydrogen peroxide concentration in human cells. More than 98% of blood catalase is localized in erythtocytes.

These cells, with their high catalase level, provide a general protection against the toxic concentration of this small hydrogen peroxide molecule. The deficiency of catalase could cause acatalasemia. Several methods have been developed for the immobilization of catalase. Immobilization is often accompanied by changes in the enzymatic activity, optimum pH, affinity to the substrate, and stability. The extent of these changes depends on the enzyme, carrier support, and the immobilization conditions. The shift in the optimum pH, from acidic or alkaline to neutral pHs, may be useful for biomedical applications since it will allow the use of some enzymes (more active at low or high pHs) under more physiological conditions. The approach developed by Sakiyama-Elbert *et al.*, consisting in a cell-triggered growth factor delivery system, may also be used for the release of other important therapeutic molecules.

APPLICATIONS OF IMMOBILIZED BIOCATALYSTS OF ENZYME IMMOBILIZATION TECHNOLOGY

At present, applications of immobilized biocatalysts include the production of useful compounds by stereospecific or regiospecific bioconversion, the production of energy by biological processes, the selective treatment of specific pollutants to solve environmental problems, continuous analyses of compounds with a high sensitivity and specificity, and medical uses such as new types of drugs for enzyme therapy or artificial organs. Immobilized enzymes are already being used in medical applications for clinical diagnosis and also for intra- and extracorporeal enzyme therapy. Applications in clinical analysis are mainly related to biosensors, which have been used to detect the presence of various organic compounds for many years. For example, glucose oxidase and catalase have been used to measure blood glucose concentration, and cholesterol oxidase and cholesterol esterase to determine cholesterol levels. In addition, enzymes can be immobilized on different prosthetic devices or used extracorporeally (*e.g.*, artificial heart, artificial lung, artificial kidney, equipment for hemodialysis and specific blood purification) as surface modifiers in order to increase the biocompatibility of these devices and to prevent blood clotting.

METHODS FOR IMMOBILIZING ENZYMES

Various methods have been developed for the immobilization of biocatalysts, which are being used extensively today. A wide range of support materials has also been employed for enzyme immobilization. The support type can be classified according to their chemical composition, such as organic or inorganic supports, and the former can be further classified into natural or synthetic matrices. Immobilization techniques can be divided into different categories: physical, chemical, enzymatic, and genetic engineering methods.

Adsorption of an enzyme

The adsorption of an enzyme onto a support or film material is the simplest method of obtaining an immobilized enzyme. Basically, the enzyme is attached to the support material by noncovalent linkages and does not require any preactivation step of the support. The interactions formed between the enzyme and the support material will be dependent on the existing surface chemistry of the support and on the type of amino acids exposed at the surface of the enzyme molecule.

Enzyme immobilization by adsorption involves, normally, weak interactions between the support and the enzyme such as ionic or hydrophobic interactions, hydrogen bonding, and van der Waals forces. Most of the support materials available have sufficient surface-charge properties suitable for immobilization by adsorption. They include inorganic carriers (ceramic, alumina, activated carbon, kaolinite, bentonite, porous glass), organic synthetic carriers (nylon, polystyrene), and natural organic carriers (chitosan, dextran, gelatin, cellulose, starch).

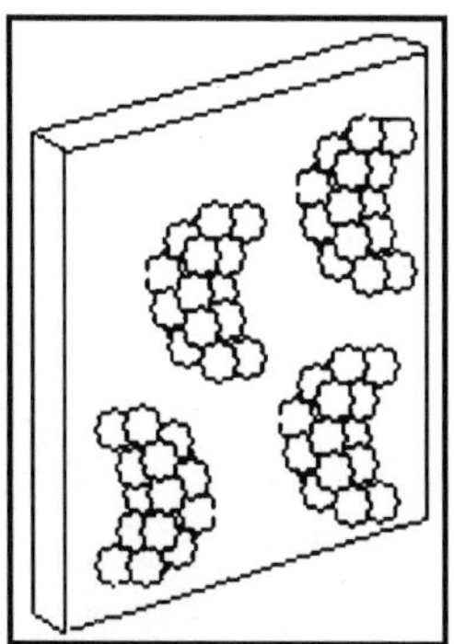

Fig. Biocatalysts Bound to a Carrier by Adsorption.

The method consists of simply mixing an aqueous solution of enzyme with the support material for a period of time, after which the excess enzyme is washed away from the immobilized enzyme on the support. The procedure requires strict control of the pH and ionic strength, because these can alter the charges of the enzyme and the support and therefore affect the level of adsorption. A simple shift in pH can cancel ionic interactions and promote the release of the enzyme from the support.

The main advantages of adsorption are the method simplicity, the little effect on the conformation/activity of the biocatalyst, and the possibility of regenerating inactive enzyme by addition of fresh enzyme. The main disadvantage is the desorption of the biocatalyst from the support due to the weak interactions established.

The enzyme desorption can easily occur by changes in the environment medium such as pH, temperature, solvent, and ionic strength or in the case of extended reactions.

Ionic Binding

Immobilization via ionic binding is based, mainly, on ionic binding of enzyme molecules or active molecule to solid supports containing ionic charges. In this method, the amount of enzyme bound to the carrier and the activity after immobilization depends on the nature of the carrier. How the enzyme is bound to the carrier. In some cases, physical adsorption may also take place. The main difference between ionic binding and physical adsorption is the strength of the interaction, which is much stronger for ionic binding, although less strong than covalent binding. The preparation of immobilized enzymes using ionic binding is based on the same procedure as described for physical adsorption.

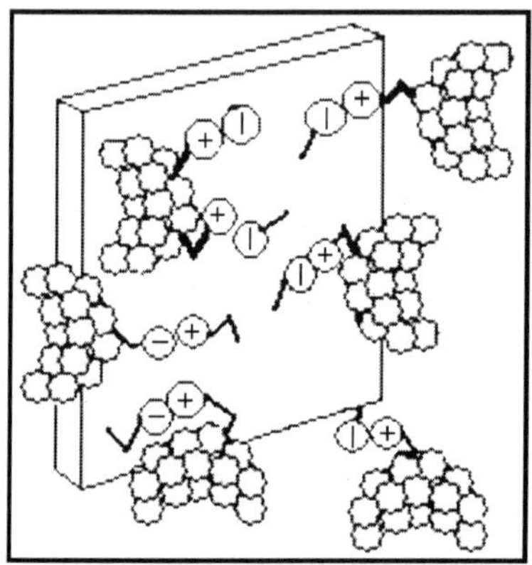

Fig. Biocatalysts Bound to a Carrier by Ionic Binding.

The ionic nature of the binding forces between the enzyme and the support also depends on pH variations, support charge, enzyme concentrations, and temperature. The supports used for ionic binding may be based on polysaccharide derivatives (*e.g.*, diethylamino-ethylcellulose, dextran, chitosan, carboxymethyl-cellulose), synthetic polymers (*e.g.*, polystyrene derivatives, polyethylene vinylalcohol), and inorganic materials (*e.g.*, ambertite, alumina, silicates, bentonite, sepiolite, silica gel, etc.). The immobilization by ionic binding has the advantage that changes in the enzyme conformation only occur in a small extent, resulting in immobilized enzymes with high enzymatic activities. The main disadvantage is the possible interference of other ions, and special attention should be paid in maintaining the correct ionic strength and pH conditions in order to prevent their easy detachment.

Covalent Binding by Chemical Coupling

The covalent binding method is based on the binding of enzymes, or other active molecules, to a support or matrix by means of covalent bonds. The bond is normally formed between a functional group present on the support surface and amino acid residues on the surface of the enzyme. Those which are most often involved in covalent binding are the amino (NH_2) group of lysine or arginine, carboxyl (CO_2H) group of aspartic acid or glutamic acid, hydroxyl (OH) group of serine or threonine, and sulphydryl (SH) group of cysteine. There are many reaction procedures for joining an enzyme to a material with a covalent bond (diazotation, amino bond, Schiff's base formation, amidation reactions, thiol-disulfide, peptide bond, and alkylation reactions). The connection between the support and the biocatalyst can be achieved either by direct linkage between the components or via an intercalated link of different length, the so-called spacer or harm. The advantage of using a spacer molecule is that it gives a greater degree of mobility to the coupled biocatalysts so that its activity can, under certain circumstances, be higher than if it is bound directly to the support. It is important to choose a method that will not involve the reaction with the amino acids present in the active site, since this could inactivate the enzyme. Basically, two steps are involved in the covalent binding of enzymes to a support material.

First, functional groups on the support material are activated by specific reagents (*e.g.*, cyanogen bromide, carbodiimide, aminoalkylethoxysilane, isothiocyanate, and epichlorohydrin, etc.). A large range of support materials is available for covalent binding, and this extensive range reflects the fact that no ideal matrix exists. Therefore, the advantages and disadvantages of a given matrix must be taken into account when considering the appropriate procedure for a given enzyme immobilization. Immobilization of enzymes through covalent attachment has also been demonstrated to induce higher resistance to temperature, denaturants, and organic solvents in several cases.

The extent of these improvements may depend on other conditions of the system, *e.g.*, the nature of the enzyme, type of support, and the method of immobilization. Many factors may influence the selection of the support, and some of the more important are its cost and availability, the binding capacity (amount of enzyme bound per given weight of matrix), hydrophilicity (the ability to incorporate water into the matrix and stability of matrix), structural rigidity, and durability during applications.

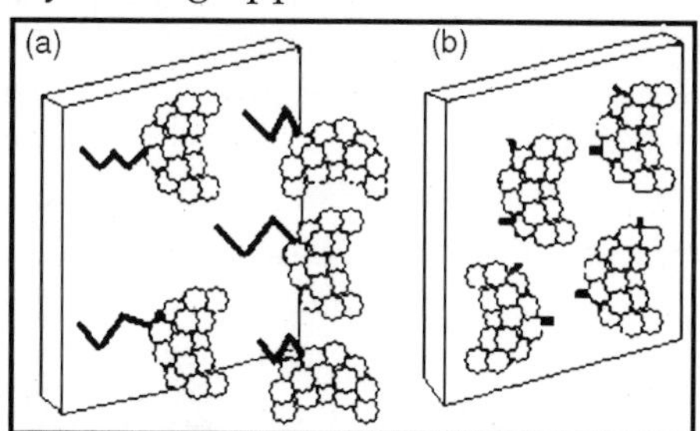

Fig. Covalent Bond Between the Biocatalysts and a Carrier with (a) and without Spacer (b).

Natural polymers, which are very hydrophilic, are popular support materials for enzyme immobilization since the residues in these polymers contain hydroxyl groups, which are ideal functional groups for participating in covalent bonds. A frequently encountered disadvantage of immobilization by covalent binding is that it places great stress on the enzyme. The necessary harshness of the immobilization procedure nearly always leads to considerable changes in conformation and a resultant loss of catalytic activity.

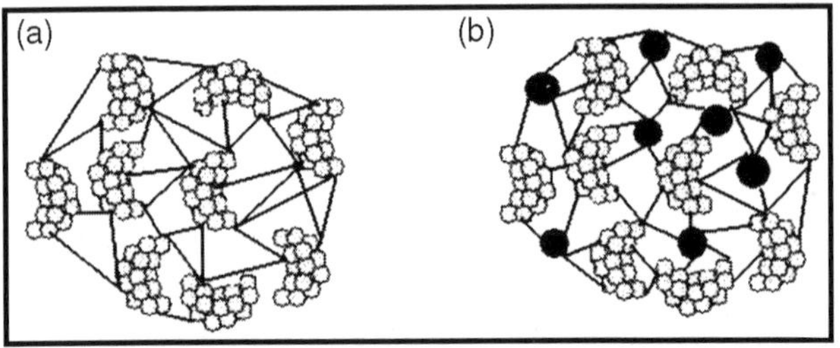

Fig. Biocatalysts Immobilized by Means of Crosslinking (a) and Co-Crosslinking with Inert Molecules Incorporated (b).

Crosslinking

The crosslinking method is based on the formation of covalent bonds between the enzyme or active molecules, by means of bi- or multifunctional reagents. The individual biocatalytic units (enzymes, organelles, whole cells) are joined to one another with the help of bi- or multifunctional reagents (*e.g.*, glutardialdehyde, glutaraldehyde, glyoxal, diisocyanates, hexamethylene diisocyanate, toluene diisocyanate, etc.). Enzyme crosslinking involves normally the amino groups of the lysine but, in occasional cases, the sulfhydryl groups of cysteine, phenolic OH groups of tyrosine, or the imidazol group of histidine can also be used for binding. How the biocatalysts can be linked by a simple crosslinking process and also by co-crosslinking, in which inert molecules are incorporated in the high-polymer network in order to improve the mechanical and enzymatic immobilized preparation. The advantages and disadvantages of a given matrix must be taken into account when considering the appropriate procedure for a given enzyme immobilization. One advantage is the simplicity of the process. The main disadvantages are the fragility of the particles produced in some cases and diffusion limitations. Since crosslinking and co-crosslinking usually involve covalent bonds, immobilized biocatalysts in this way frequently undergo changes in the conformation with a resultant loss of activity. The isomerization of glucose process is a very important example of the industrial application using biocatalysts crosslinked with glutaraldehyde. Some of the immobilized preparations used in these large-scale processes are produced simply by glutaraldehyde treatment of bacterial cell masses that have formed fine particles.

Entrapment and Encapsulation

The entrapment method for immobilization consists of the physical trapping of the active components into a film, gel, fibre, coating, or microencapsulation. This method can be achieved by mixing an enzyme or active molecule with a polymer and then crosslinking the polymer to form a lattice structure that traps the enzyme. Microencapsulated enzymes are formed by enclosing enzymes solution within spherical semipermeable polymer membranes with controlled porosity.

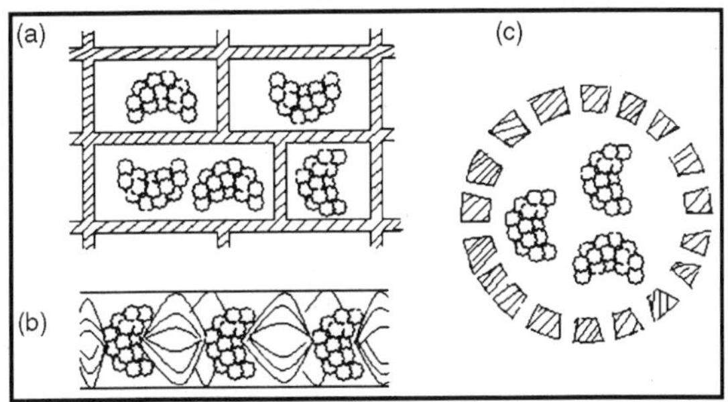

Fig. Enzyme Encapsulation in a Matrix (a), Fibre (b), or Capsule (c).

While the encapsulation of dyes, drugs, and other chemicals has been known for some time, it was not until the mid-1960s that such a method was first applied to enzymes.

Since that first report, a number of other enzymes have been successfully immobilized via microencapsulation, using a number of different materials and methods to prepare the microcapsules. The advantages of this immobilization method are the extremely large surface area between the substrate and the enzyme, within a relatively small volume, and the real possibility of simultaneous immobilization. The major disadvantages of this method include the occasional inactivation of enzyme during microencapsulation and the high enzyme concentration required. In addition, to retain the enzyme, the pore size needs to be very low and these systems tend to be very diffusion limited.

Protein Fusion to Affinity Ligands and Enzymatic Conjugation

As described before, there are many methods for protein immobilization, but some of them require chemical modification of the matrix, which may result in material degradation, especially when biodegradable polymers are used. In addition, these modifications, necessary to attach the enzyme to the matrix, often result in the loss of enzyme activity as well as the inclusion of toxic organic compounds, which have to be removed before the system can be used in biomedical applications.

In this type of application, the efficacy of immobilized biomolecules for stimulating specific cell responses (*e.g.*, proliferation or differentiation) depends on the mode by which these modulators are presented to the target cells. In these cases, it is important to ensure the correct orientation and full bioactivity of the molecules when they are immobilized. Covalent binding by chemical coupling, again, might hinder ligand–receptor interaction or prevent receptor dimerization and capping on target cells. In nature, there are certain protein molecules, such as lectins, avidin, immunoglobulin G (IgG) binding domains of protein G and protein A, and carbohydrate binding modules (present in many polysaccharide-degrading enzymes), that bind with high affinity and specifically to certain molecules or solid surfaces. These binding domains may be used as affinity tags for immobilizing proteins to affinity adsorbents.

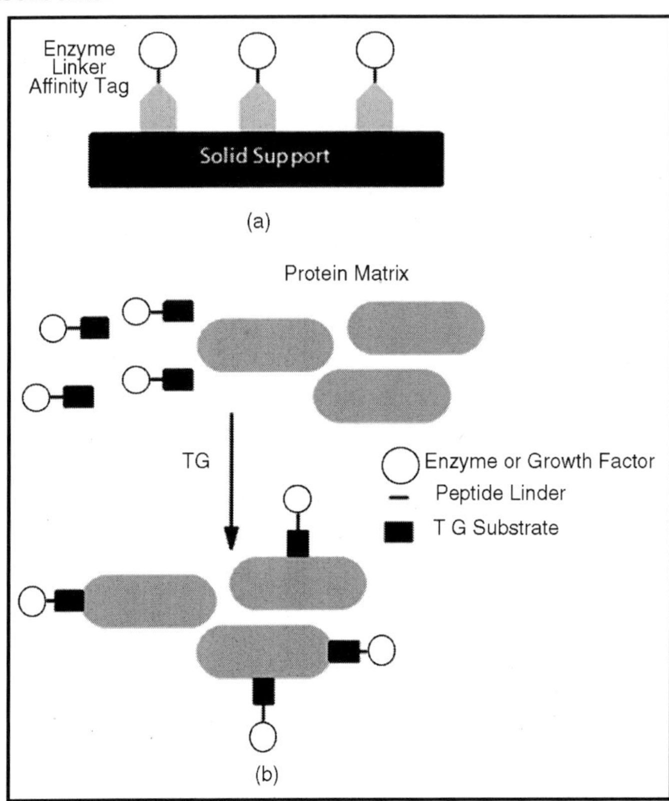

Fig. Enzyme Immobilization to Solid Matrices Via Protein Fusion to an Affinity Ligand (a) and Enzymatic Conjugation Catalyzed by Transglutaminase (b).

In this technique, DNA encoding a polypeptide affinity tag is fused to the gene of interest, and the expression of the gene results in a fusion protein. Such a fusion protein could be immobilized by the specific binding of the affinity tag to an affinity adsorbent.

With this method, the conformational changes in the protein upon immobilization are minimal, and the immobilized biomolecule could retain high activity. In addition, since fusion proteins are specifically immobilized on the support materials, these methods normally allow the immobilization of high densities of ligands and can also simplify the immobilization procedure.

GLUCOSE-6-PHOSPHATE DEHYDROGENASE DEFICIENCY

Glucose-6-phosphate dehydrogenase (G6PDH) is expressed in all cells and is responsible for the first reaction of the pentose phosphate pathway in which glucose-6-phosphate is oxidized to 6-phosphogluconolactone with concomitant production of NADPH. he NADPH produced via the pentose phosphate pathway is required for a variety of reductive biosynthetic reactions as well as for the regeneration of the reduced form of glutathione (GSH). GSH is essential for the detoxification of hydrogen peroxide (H_2O_2) and therefore, cellular defence against the oxidizing effects of H_2O_2 is absolutely dependent upon the generation of NADPH. GSH converts H_2O_2 to H_2O via the action of glutathione peroxidase and requires 1 mole of NADPH per mole of H_2O_2. The critical need for production of NADPH via the pentose phosphate pathway is especially true in red blood cells because there are no other NADPH-producing reactions in these cells and they are extremely sensitive to oxidative damage.

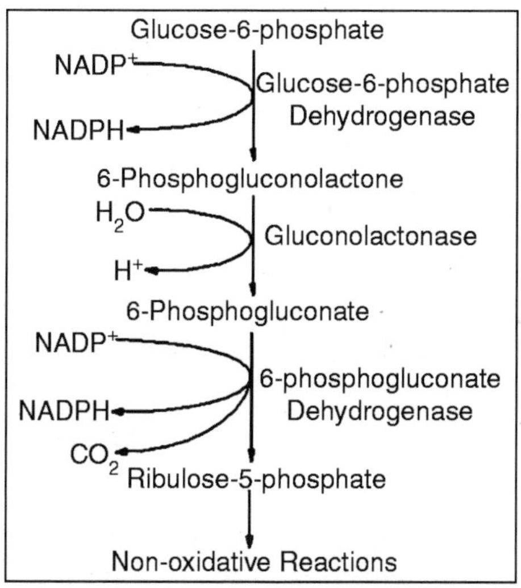

Fig. Oxidative Stage of Pentose Phosphate Pathway

The glucose-6-phosphate dehydrogenase gene (symbol G6PD) is located on the X chromosome (Xq28) about 1 Mb (million base pairs) from the telomeric end. The gene spans 18 kb and is composed of 13 exons encoding a protein of 531 amino acids. The first exon of the gene is a non-coding exon with the translational initiation codon present in exon 2. Post-translational processing G6PDH results in a 515 amino acid functional protein containing an acetylated alanine residue at the N-terminus. Biologically active G6PDH is functional as either a homodimer or a homotetramer and both forms co-exist in equal proportions at neutral pH.

Deficiencies in G6PDH are the most commonly inherited enzyme deficiencies (enzymopathies) world wide with estimates of greater than 400 million affected individuals. The incidence of deficiency approaches 25% of the population of persons of Mediterranean, tropical African, and tropical and subtropical Asian descent. In fact, the incidence of G6PDH deficiency is so high in some populations that the occurrence of homozygous females is not at all rare as would normally be expected for an X-linked disorder. The inheritance of G6PDH deficiency is clinically classified as an X-linked recessive disorder, however, because heterozygous females can develop hemolytic episodes the disorder is not truly recessive in the Mendelian sense. The phenomenon of symptom manifesting heterozygous females is explained by the co-existence of populations of G6PDH positive and negative cells in the same female as a result of X-chromosome inactivation.

Deficiency in G6PDH represents one of the most genetically heterogeneous disorders. There are over 400 different variants of G6PDH defined by their diverse biochemical characteristics. To date a total of 130 different point mutations have been identified in the G6PD gene. Surprisingly, only five in-frame deletions have been identified and no large deletions or insertions have been found. This fact is likely explained by the observation that in mouse models of G6PDH deficiency, complete loss of enzyme activity is associated with embryonic lethality.

G6PDH deficiency cannot be classified by a single mutation but is manifest as a consequence of numerous structural allelic mutants. In addition, G6PDH deficiencies are also classifiable by a variety of physicochemical parametres including chromatographic properties, thermostability, pH dependence, K_M for glucose-6-phosphate, and K_M for NADP. G6PD gene variants are currently divided into five classifications dependent upon enzyme activity and clinical manifestations. In addition, the deficiencies are classified as to whether or not they are sporadic or polymorphic.

The five classifications are:

Class 1: Enzyme deficiency with chronic nonspherocytic hemolytic anemia

Class 2: Severe enzyme deficiency, less than 10% of normal activity

Class 3: Moderate to mild enzyme deficiency, 10-60% of normal activity

Class 4: Very mild or no enzyme deficiency, at least 60% of normal activity

Class 5: Increased enzyme activity.

Mutations in the G6PD gene that result in nonspherocytic hemolytic anemia have all been found to cluster near the carboxy terminal end of the protein, whereas, the mutations that result in clinically mild symptoms are clustered near the amino terminal end of the protein. Nearly all G6PDH mutations are single nucleotide missense mutations resulting in the substitution of a single amino acid. In most instances of G6PDH mutation resulting in disease it is due to protein instability as a result of the amino acid substitution.

Clinical Features of G6PDH Deficiency

The vast majority of individuals that harbor a mutant form of the G6PD gene go their entire lives without knowing they carry a mutation. The acute hemolysis that is the only common clinical manifestation in G6PD mutations can rapidly be compensated for and thus may remain undetectable. The most common symptoms of G6PDH deficiency are neonatal jaundice and acute hemolytic anemia.

In addition, red cell deficiency in G6PDH is the basis of favism, primaquine sensitivity (a drug used in the treatment of malaria similar to quinine) and some other drug-sensitive hemolytic anemias (*e.g.* due to sulfonamides used to treat bacterial infections).

Favism is a term relating to a hemolytic anemia that results from the ingestion of the fava bean (*Vicia faba*) also known as broad beans. In the case of primaquine-induced hemolytic anemia, it was this clinical manifestation, particularly in Blacks, that actually led to the discovery of G6PDH deficiencies. Aside from favism and drug-induction of hemolytic anemia, infection is the most common cause of hemolysis in G6PDH deficient individuals. With respect to favism, the occurrence of acute hemolysis after ingestion of the fava bean was recognized as far back as the time of the Greek mathematician, Pythagoras.

Analysis of extracts from fava beans demonstrated that the toxic compounds are the pyrimidine aglycones divicine and isouramil in combination with ascorbic acid.

LACTATE DEHYDROGENASE

Lactate dehydrogenase (LDH or LD) is an enzyme present in a wide variety of organisms, including plants and animals. Lactate dehydrogenases exist in four distinct enzyme classes.

Two of them are cytochrome c-dependent enzymes with each acting on either D-lactate or L-lactate. The other two are NAD(P)-dependent enzymes with each acting on either D-lactate or L-lactate. This article is about the NAD(P)-dependent L-lactate dehydrogenase.

REACTIONS

Lactate dehydrogenase catalyzes the interconversion of pyruvate and lactate with concomitant interconversion of NADH and NAD^+. It converts pyruvate, the final product of glycolysis to lactate when oxygen is absent or in short supply, and it performs the reverse reaction during the Cori cycle in the liver.

At high concentrations of lactate, the enzyme exhibits feedback inhibition and the rate of conversion of pyruvate to lactate is decreased. It also catalyzes the dehydrogenation of 2-Hydroxybutyrate, but it is a much poorer substrate than lactate. There is little to no activity with beta-hydroxybutyrate.

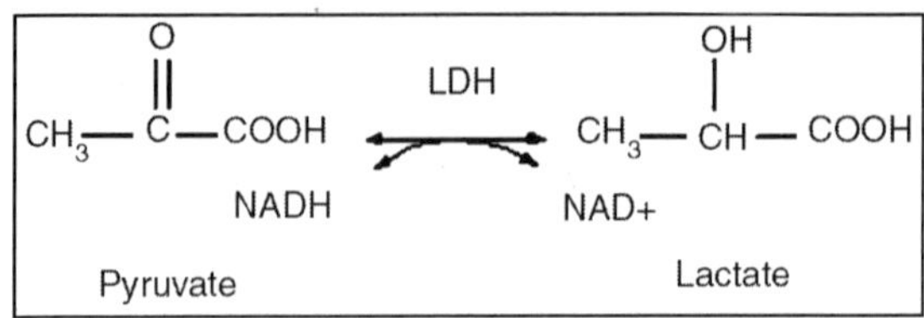

Fig. Catalytic Function of LDH

Enzyme Isoforms

Functional lactate dehydrogenase are homo or hetero tetramers composed of M and H protein subunits encoded by the LDHA *and* LDHB *genes respectively:*

- LDH-1 (4H) - in the heart and RBCs
- LDH-2 (3H1M) - in the reticuloendothelial system
- LDH-3 (2H2M) - in the lungs
- LDH-4 (1H3M) - in the kidneys, placenta and pancreas
- LDH-5 (4M) - in the liver and striated muscle

The five isozymes that are usually described in the literature each contain four subunits. The major isozymes of skeletal muscle and liver, M_4, has four muscle (M) subunits; while H_4 is the main isozymes for heart muscle in most species, containing four heart (H) subunits. The other variants contain both types of subunits.

Usually LDH-2 is the predominant form in the serum. A LDH-1 level higher than the LDH-2 level (a "flipped pattern"), suggests myocardial infarction (damage to heart tissues releases heart LDH, which is rich in LDH-1, into the bloodstream). The use of this phenomenon to diagnose infarction has been largely superseded by the use of Troponin I or T measurement.

Genetics in Humans

The M and H subunits are encoded by three different genes:

- The M subunit is encoded by *LDHA,* located on chromosome 11p15.4 (Online 'Mendelian Inheritance in Man' (OMIM) 150000)
- The H subunit is encoded by *LDHB,* located on chromosome 12p12.2-p12.1 (Online 'Mendelian Inheritance in Man' (OMIM) 150100)
- A third isoform, *LDHC* or *LDHX,* is expressed only in the testis (Online 'Mendelian Inheritance in Man' (OMIM) 150150); its gene is likely a duplicate of *LDHA* and is also located on the eleventh chromosome (11p15.5-p15.3)

Mutations of the M subunit have been linked to the rare disease *exertional myoglobinuria* and mutations of the H subunit have been described but do not appear to lead to disease.

Medical use

Tissue breakdown elevates levels of LDH, and therefore a measure of it indicates, *e.g.*, hemolysis. Other disorders indicated by elevated LDH include cancer, meningitis, encephalitis, acute pancreatitis, and HIV.

Hemolysis

In medicine, LDH is often used as a marker of tissue breakdown as LDH is abundant in red blood cells and can function as a marker for hemolysis. A blood sample that has been handled incorrectly can show false-positively high levels of LDH due to erythrocyte damage. It can also be used as a marker of myocardial infarction. Following a myocardial infarction, levels of LDH peak at 3-4 days and remain elevated for up to 10 days. In this way, elevated levels of LDH (where the level of LDH1 is higher than that of LDH2) can be useful for determining if a patient has had a myocardial infarction if they come to doctors several days after an episode of chest pain.

Tissue Turnover

Other uses are assessment of tissue breakdown in general; this is possible when there are no other indicators of hemolysis. It is used to follow-up cancer (especially lymphoma) patients, as cancer cells have a high rate of turnover with destroyed cells leading to an elevated LDH activity.

Exudates and Transudates

Measuring LDH in fluid aspirated from a pleural effusion (or pericardial effusion) can help in the distinction between exudates (actively secreted fluid, *e.g.* due to inflammation) or transudates (passively secreted fluid, due to a high hydrostatic pressure or a low oncotic pressure). The usual criterion is that a ratio of fluid LDH versus upper limit of normal serum LDH of more than 0.6 or T! indicates an exudate, while a ratio of less indicates a transudate. Different laboratories have different values for the upper limit of serum LDH, but examples include 200 and 300 IU/L. In empyema, the LDH levels, in general, will exceed 1000 IU/L.

Meningitis and Encephalitis

High levels of lactate dehydrogenase in cerebrospinal fluid are often associated with bacterial meningitis. In the case of viral meningitis, high LDH in general indicates the presence of encephalitis and poor prognosis.

HIV

LDH is often measured in HIV patients as a non-specific marker for pneumonia due to *Pneumocystis jiroveci* (PCP). Elevated LDH in the setting of upper respiratory symptoms in an HIV patient suggests, but is not diagnostic for, PCP. However, in HIV-positive patients with respiratory symptoms, a very high LDH level (>600 IU/L) indicated histoplasmosis (9.33 more likely) in a study of 120 PCP and 30 histoplasmosis patients.

5

Enzymes in Industry

ENZYMES FOR INDUSTRIAL APPLICATIONS

STARCH-PROCESSING INDUSTRIES

Specialty Enzymes' SEBstar-HTL and SEBamyl-BAL are blends used in the starch-liquefaction process for production of maltodextrins that are used in paperboard adhesives, paper sizing and in the manufacture of syrup-sweeteners.

Fungal alpha-amylase SEBamyl-L is used to saccharify low-DE, starch hydrolyzates in the production of crystalline maltose, low-DE maltose syrups and high-conversion, mixed maltose/glucose syrups. These stable sugar-syrups resist crystallization. They are also widely used as sweeteners in the beverage and food industries and as a source of fermentable sugars in the brewing and fermentation industries.

Detergents

Specialty Enzymes products, SEBrite-BP are a range of alkaline proteases that when added to surfactant-containing liquid and powder laundry detergents or pre-spotters, enhance the removal of protein-containing stains such as milk, egg, soya, blood, grass and body fluids of human origin. The use of Specialty Enzymes detergent alpha-amylases enzymes such as SEBrite-BA and SEBrite-A in liquid and powder laundry detergents solubilizes starch-containing stains such as baby food, spaghetti, mashed potatoes, oatmeal, gravy, chocolate, tomato sauce and other starch thickened foods. To enhance stain removal, brighten the colour, and soften cotton and cotton-containing garments, Specialty Enzymes offers a laundry-detergent compatible cellulase enzyme, SEBrite-COLOUR. SEBrite-COLOUR, effectively removes cotton cellulose microfibrils on wear and wash-damaged cotton garments. Repeated washing of cotton-containing garments with a laundry detergent containing SEBrite-COLOUR, improves soil removal, brightens fabric colour, softens and restores original garment appearance. Specialty Enzymes' protease and amylase enzymes, SEBalase-BP and SEBrite-BA can be used cost-effectively to improve the wash performance of reduced-pH, low temperature industrial and institutional laundry detergents used to remove blood, body fluids, organic material and food soils from hospital, restaurant and slaughterhouse textiles. Specialty Enzymes protease enzymes SEB-Prolase P or SEBalase-PB, carbohydrase enzyme, CelluSEB-T, alpha-amylase enzyme SEBamyl-B, Spectinase enzyme, ExtractSEB-R and lipase enzyme, SEBrite-L can be used to effectively clean filter membranes fouled with protein, starches, fats, gums, cellulose and pectin residues.

Leather

Specialty Enzymes' SEBsoak product, used during the soaking process, removes unwanted parts of hides and skins, provides an uniform soaking effect, speeds up the soaking process and better re-hydrates hides previously soaked in brine.

SEBlime is a blend that loosens hair from the corium layer of the skin, making hair removal easy. This method is faster, safer and more convenient than the method using sulphide. SEBlime offers eco-friendly, biodegradable, sulfide-free liming. It has optimum swell regulating properties, which result in excellent grain smoothness.

SEBbate Acid improves the hide's pliability and makes the grain smoother and more uniform for fabrication and dyeing. SEBbate Alkali helps stabilize the hide by ensuring the removal of all non-leather forming proteious matter.

Paper and Pulp

Specialty Enzymes' SEBrite-BB is custom made for bleach boosting in paper/wood pulping applications. It reduces pulp bleaching chemical consumption and effectively boosts the performance of kraft pulp bleaching systems.

Personal Care

Our special combinations of enzymes like SEBCareHR 413 L, SEBCareHR 413 P, SEBCareHR 414 L and SEBCareHR 414 P can be used to remove contaminants like dust, dead cells, and excessive oil, which often clog the hair follicle openings. Using our special blends in hair care formulations can clean the scalp, thereby increasing blood circulation, and leading to healthy keratin formation and less hair loss. Specialty Enzymes offers SEBCareOR 115L, which can be added to oral care products to break down excessive sugar in the mouth, thus preventing tooth decay, odour and discolouration. Specialty Enzymes also has proteolytic enzyme blends to break down the protein build up on teeth and gums - SEBCareOR 104 B, SEBCareOR 114 L and SEBCareOR 114 P. These are special blends of proteolytic enzymes, which can be used to prevent plaque development on teeth and gums and help whiten teeth. Our proteolytic enzyme blends like SoftSEB-SK 3131 L, SoftSEB-SK 3131 P, SoftSEB-

SK 320 L and SoftSEB-SK 320 P, further help in this process by breaking down the keratin in dead cells, thus gently peeling off the dead outer layer without damaging the inner living cells. Our other blends, like SEBCareSK 309 L, are used in skin care products to remove excessive oil on skin, leaving it non-oily, soft and silky.

Pharmaceutical

Specialty Enzymes' bacterial blend Peptizyme SP is being used as a highly effective alternative to antibiotics in many countries. It treats inflammatory disorders, by not only fighting inflammation, but also relieving pain and swelling, improving recovery time and stimulating the immune system. Peptizyme SP has a "scavenging" effect. It helps chelate the heavy metals through which the body release toxins, and hence modulates the immune system, addresses hormonal imbalances and speeds wound and tissue repair time.

Textiles

Specialty Enzymes' biopolishers, such as Denisoft and Denistone PW help manufacturers achieve a permanent soft feel, increased absorbency and cool breathing nature of the fabric. More and more manufacturers are reducing the use of chemicals, like cationic softeners and silicones, and opting for enzyme-based biopolishing agents, particularly due to their ability to obtain permanent softness.

ENZYMES USED IN THE DAIRY INDUSTRY

In the field of biotechnology there are many food science applications that utilize enzymes. In the dairy industry, some enzymes are required for the production of cheeses, yogurt and other dairy products, while others are used in a more specialized fashion to improve texture or flavour.

Five of the more common types of enzymes and their role in the dairy industry are described below:

1. *Rennet*: Milk contains proteins, specifically caseins, that maintain its liquid form. Proteases are enzymes that are added to milk during cheese production, to hydrolyze caseins, specifically kappa casein, which stabilizes micelle formation preventing coagulation. Rennet and rennin are general terms for any enzyme used to coagulate milk.
2. *Other Proteases*: Milk contains a number of different types of proteins, in addition to the caseins. Cow milk also contains whey proteins such as lactalbumin and lactoglobulin. The denaturing of these whey proteins, using proteases, results in a creamier yogurt product. Destruction of why proteins is also essential for cheese production.
3. *Lactase*: Lactase is a glycoside hydrolase enzyme that cuts lactose into it's constituent sugars, galactose and glucose. Without sufficient production of lactase enzyme in the small intestine, humans become

lactose intolerant, resulting in discomfort (cramps, gas and diarrhea) in the digestive tract upon injestion of milk products. Lactase is used commercially to prepare lactose-free products, particularly milk, for such individuals. It is also used in preparation of ice cream, to make a creamier and sweeter-tasting product. Lactase is usually prepared from *Kluyveromyces sp.* of yeast and *Aspergillus sp.* of fungi.

4. *Catalase*: The enzyme Catalase has found limited use in one particular area of cheese production. Hydrogen peroxide is a potent oxidizer and toxic to cells. It is used instead of pasteurization, when making certain cheeses such as Swiss, in order to preserve natural milk enzymes that are beneficial to the end product and flavour development of the cheese. These enzymes would be destroyed by the high heat of pasteurization. However, residues of hydrogen peroxide in the milk will inhibit the bacterial cultures that are required for the actual cheese production, so all traces of it must be removed. Catalase enzymes are typically obtained from bovine livers or microbial sources, and are added to convert the hydrogen peroxide to water and molecular oxygen.
5. *Lipases*: Lipases are used to break down milk fats and give characteristic flavours to cheeses. Stronger flavoured cheeses, for example, the italian cheese, Romano, are prepared using lipases. The flavour comes from the free fatty acids produced when milk fats are hydrolized. Animal lipases are obtained from kid, calf and lamb, while microbial lipase is derived by fermentation with the fungal species *Mucor meihei*.

DAIRY PRODUCTS

Throughout the first half of this century milk and products derived from it remained of major importance. At mid-century world milk production was around 180 million tons annually, and from this were made some three million tons of butter and two million tons of cheese. Most countries remained more or less self-supporting in respect of these commodities: few imported more than 10 per cent of their over-all needs. The position of the United Kingdom was quite different, however: in 1939 she imported 80 per cent of her butter and 50 per cent of her cheese. The main exporters were New Zealand, by far the largest, followed by Australia, Denmark, and Holland. While technological developments, notably mechanization of existing methods and centralization of manufacture, ensured that quantities were sufficient, this was not always the case so far as quality was concerned. Factory-made butter was the least affected, in that the process of its manufacture is relatively simple. With cheese, however, the decline was very evident, and mainly the result of commercial factors. The full flavour of a cheese is realized only as the result of a long ripening process governed by a complex system of enzymes. If this process is curtailed, the product, though unimpaired from the nutritional point of view, may be relatively tasteless and of poor texture; such curtailment did occur in the cheese factories and stores, though it would be idle to suggest that all farmhouse cheese was perfect.

Much of the decline was promoted by changing demand. Bread and cheese was the traditional meal of the industrial worker in the factory and the demand was for quantity rather than quality. To satisfy this, quickripening moist cheeses were produced on the basis of rapid turnover and larger profits. Less cheese was stored in lofts, to ripen, but more in cold-stores where microbiological action was suspended. Worse, means were discovered of making cheese from pasteurized milk.

At first the process denatured milk proteins and prevented coagulation with rennet, but this was overcome by using a somewhat lower pasteurization temperature. The result was 'processed' cheese, first manufactured in Switzerland at the beginning of the century and later made elsewhere in Europe and in North America. From the purely practical point of view it has much to commend it: although its moisture content is higher than that of real cheese, thus making it poorer value for money, it keeps indefinitely and there is no wasteful rind. Its disadvantage is its taste and texture.

We have already noted the advent of margarine as a rival to butter. The original product was in effect little more than an emulsion of purified beef tallow with milk, from which the fatty fraction was then precipitated by addition of ice and worked to a buttery consistency. It was not very palatable but it was cheap and nutritious. In the twentieth century vegetable fats and oils were introduced and the skim milk used was subjected to lactic fermentation.

Great improvements were made in the emulsification process, and incorporation of vitamins A and D (compulsory in Britain during the Second World War) was common. By 1950 the palatability of margarine and its nutritional value had been greatly enhanced. By careful control of the manufacturing process, products could even be made suitable for different climatic conditions. World production in 1938 amounted to more than one million tons, roughly one-third that of butter.

Milk contains approximately 87 per cent water; it is not surprising therefore that there has long been an interest in drying (or condensing) it, not only as a means of indefinite preservation but also to reduce its bulk. Most early processes depended on direct heating: the milk was allowed to run over a hot roller and the dry residue was then scraped off. Unfortunately this causes considerable change in flavour and the lost water is not easily reabsorbed when desired. Nevertheless, it provides a product useful for various purposes in the food industry.

During the Second World War spraydrying processes were developed in which a spray of skimmed milk was allowed to fall through a rising current of hot air; by the time the droplets reached the bottom of the plant they have turned to powder. Again, however, there was difficulty in reconstituting the product. In 1946 this was overcome by an agglomeration process developed in the USA by the Instant Milk Company. Spray-drying was also used to dry eggs.

There were some important changes in the process of pasteurization. From the 1920s this was done in bulk rather than in bottles and consisted in holding the milk for half an hour at about 65°C; it was then rapidly cooled. During the Second World War, however, this method began to be replaced by the so-called HTST (High Temperature Short Time) process in which the working temperature was about 72°C but the operating time only a quarter of a minute. It was followed by the UHT (Ultra High Temperature) process in which the milk is superheated to about 150°C for only a few seconds. The short heating periods of these last two processes demanded the development of elaborate heat exchangers.

ULTRA-PASTEURIZED MILK

Today, an increasing amount of milk found in conventional grocery stores—including most organic milk—is ultra-pasteurized. The official U.S. government definition of an ultra-pasteurized dairy product stipulates "such product shall have been thermally processed at or above 280° F for at least 2 seconds, either before or after packaging, so as to produce a product which has an extended shelf life under refrigerated conditions." Confusingly, ultra-pasteurized milk is oftentimes referred to as or labeled as UHT, for "ultra-high temperature." It is the high-temperature processing that gives the milk an extended shelf life (ESL).

Why do processors embrace UHT milk? Because today's milk is no longer a local product; it is processed in huge processing plants and then shipped all over the country. When packaged in aseptic containers, UHT milk remains stable at room temperature for up to six months. Its extended shelf life with refrigeration in standard packaging, such as plastic bottles, is up to 50 days—enough time for it to be shipped across country, or internationally, and sold to customers far from the milk's origin.

HARMFUL PROCESS

In the commercial processing of UHT milk, raw milk is first preheated to 176-194° F, then submitted to one of two heating methods: direct or indirect. In the direct method, milk is injected with superheated steam or the milk is sprayed into steam. This raises the temperature of the milk immediately, but also slightly dilutes it. The extra water is removed when the milk is subsequently cooled in a vacuum chamber. Indirect heating occurs by bringing milk into contact with super-heated metal plates that have been heated by steam—hence, the steam is "indirectly" heating the milk. Some new systems combine the two processes.

According to Lee Dexter, microbiologist and owner of White Egret Farm goat dairy in Austin, Texas, ultra-pasteurization is an extremely harmful process to inflict on the fragile components of milk. Dexter explains that milk proteins are complex, three-dimensional molecules, like tinker toys. They are broken down and digested when special enzymes fit into the parts that stick out.

Rapid heat treatments like pasteurization, and especially ultra-pasteurization, actually flatten the molecules so the enzymes cannot do their work. If such proteins pass into the bloodstream (a frequent occurrence in those suffering from "leaky gut," a condition that can be brought on by drinking processed commercial milk), the body perceives them as foreign proteins and mounts an immune response. That means a chronically overstressed immune system and much less energy available for growth and repair.

RESEARCH

Scientists in Australia, a country with a huge dairy industry, have taken the lead in researching UHT milk. A 2002 paper discusses how UHT processing and subsequent storage causes several changes affecting the shelf life of UHT milk. The changes include: whey protein denaturation, protein-protein interaction, lactose-protein interaction, isomerisation of lactose, Maillard browning which imparts a burnt flavour, sulphydryl compound formation, formation of a range of carbonyl and other flavour-imparting compounds, and formation of insoluble substances.

According to the authors Datta and Deeth, these changes "ultimately reduce the quality and limit the shelf life of UHT through development of off flavors, fat separation, age gelation and sedimentation." Nevertheless, according to the report, the milk remains "commercially stable."

A thorough reading of their paper reveals a very interesting point. During the heating process, the aforementioned sulphydryl compounds impart a very strong cabbagy off-flavour to UHT milk that is most noticeable immediately after heating. These compounds dissipate during storage, but approximately one month into storage, UHT milk begins to deteriorate and is described in the industry as "stale."

In the later stages of storage, a bitter taste develops and then it undergoes "age gelation," a process in which the milk becomes more viscous and eventually loses fluidity. So, it seems the optimum time to drink UHT milk with any degree of enjoyment, if that's even possible, is limited to the interval between the dissipation of the cabbage flavour and the onset of staleness, bitterness and gelatinous conditions. In the U.S., these off-flavors seem to go unnoticed, which makes me wonder whether some kind of flavorings or other chemicals are being added to UHT milk? If the whole industry does this, they don't need to list such additives on the label because it is an "industry standard."

PACKAGING

UHT milk was introduced by the Italian dairy company Parmalat, which sold UHT milk in aseptic tetra-brik containers as a convenience food to Europeans, most of whom lived in apartments and had small refrigerators. That strategy didn't work in the U.S., where almost everyone has a large refrigerator and where consumers still value "fresh" milk.

In the early 1990s, in order to overcome American consumer resistance to milk that didn't need to be refrigerated, Parmalat implemented an aggressive marketing plan as well as a strategy to overcome "regulatory impediments," hoping to carve out a niche in the U.S. milk market. While little boxes of Parmalat UHT milk can be found in the grocery aisle, most U.S. milk producers package UHT milk in cartons and plastic bottles identical to those containing pasteurized milk. While UHT milk remains popular in Europe and Brazil, in the United States, consumer resistance has spurred Horizon and Organic Valley, the major producers of organic milk, to reintroduce pasteurized milk in some locations where they have local sourcing and nearby milk plants. In the Washington, DC area, Horizon sells UHT milk in the supermarkets and pasteurized milk in the upscale markets like Whole Foods.

PACKAGING

While the processing of UHT milk creates palatability problems and possible health risks, so does its packaging—both the aseptic boxes and plastic containers. For example, phthalates and other endocrine disrupting compounds (EDC) can leach into the milk. In a recent study published in the *Journal of Agricultural and Food Chemistry*, investigators measured the presence of nonylphenol (NP), bisphenol A (BPA) and bisphenol A diglycidyl ether (BADGE) in two brands of UHT milk in tetra-brik containers. These containers are designed to be stacked on pallets and are lined with polyethylene. They also looked at EDCs in two brands of milk that had been bottle sterilized (heated to at least 212° F) "for a length of time sufficient to render it commercially sterile" and in one brand of canned powdered infant formula. All the samples contained measurable levels of endocrine disrupting substances that leaked from the plastic of the containers, or plastic lining the containers. Even when kept cold, plastic will leach some chemicals into the liquid it contains; filling plastic-lined containers with superheated milk or subjecting liquid-filled containers to high heat is a recipe for the release of phthalates and similar substances. The researchers noted that the levels of these compounds in the samples studied did not achieve "the maximum leached level allowed by law." Their concluding comment, however, is more pessimistic: The impact these compounds may have on organisms and human beings needs to be further studied, especially with regard to accumulation, degradation and possible effects within the endocrine system."

UHT MILK FOR FUTURE

In the fall of 2004, we learned from a dairy inspector in a mid-Atlantic state that the FDA had conducted a nationwide conference call with the dairy departments of all 50 states. There were two reported topics of discussion on the agenda. The first topic concerned raising the required temperature of pasteurization. The stated reason: many organisms have become heat resistant and now survive the pasteurization process.

The Johne's, or paratuberculosis, bacterium, is a good example. Johne's disease is endemic in today's confinement dairies and has been linked to Crohn's disease in humans. Many samples of pasteurized milk now test positive for Johne's bacteria. B. cereus and botulism spores also survive, as do those of protozoan parasites.

Is the FDA planning to raise the required temperature of pasteurization to that now used in ultra-pasteurization? If so, the agency has not yet published an official plan. The 2005 "Programme Priorities" for the FDA's Centre for Food Safety and Applied Nutrition (CFSAN) include finding ways to reduce the incidence of foodborne *Listeria* outbreaks, which includes continuing to develop and deliver State and Federal curricula for "*Listeria* control implementation for manufacturers and retail/food service operators."

Whether this is a covert move that will eventually lead to raising the temperature of pasteurization remains to be seen. Such a move would redefine ultra-pasteurization as pasteurization so that the words "ultra-pasteurization" or UHT might then not have to appear on the label. The industry would certainly be happy about such a move, but would consumers? Any major move in this direction requires the FDA to keep the public informed through an Advance Notice of Proposed Rulemaking (ANPRM), which should include a period for public comment.

The second reported topic of discussion during the FDA dairy meeting was an organization called the Weston A. Price Foundation. We are definitely on their radar screen!

A CULTURE PROBLEM

In the name of science, we decided to do an experiment.usually make homemade yogurt with a Bulgarian culture and the best quality milk we can find. Raw milk is ideal, but sometimes we have to settle for pasteurized, un-homogenized milk. Both raw and pasteurized, un-homogenized milk produce a firm-textured yogurt with a delectable layer of cream on the top.

This time, we bought a quart container of organic Horizon UHT whole milk and cultured it exactly the same way. It took a few hours longer to set up and it never attained the consistency we have come to expect and enjoy. When we tried to spoon some out of the jar, it dissolved into small curds instead of staying in a firm mass on the spoon. It became very watery and unappetizing—and ended up down the sink instead of in my family's stomach.

At least pasteurized milk will make an acceptable yogurt. Milk that has been sterilized will not. we haven't noticed the UHT label on any cheese packages either. Since ultra-pasteurized or UHT milk will not adequately support microbial life, it is unlikely that it will adequately support human life either. The fact that the processing and storage of UHT milk is so horrendous makes me wonder why they bother. Fermentation is the best way to extend milk's shelf life and should be re-examined as a better alternative.

ENZYMES FOR THE ALCOHOL INDUSTRY

Specialty Enzymes and Biochemicals Co. is a worldwide leader in the production of industrial enzymes. Enzymes are eco-friendly solutions to industrial problems. They can improve manufacturing efficiency and lower costs. Specialty Enzymes and Biochemicals Co. prides itself on being one of the few manufacturers in the world that produces a full spectrum of industrial enzymes derived from all four natural origins: plant, animal, fungus and bacteria. And we use both solid state and submerge culture techniques of fermentation. Specialty Enzymes and Biochemicals Co. specializes in developing tailored enzyme solutions for individual customers. Our growing production base, and comprehensive advisory and after sales services, allow us to cater to our rapidly expanding global markets.

Products

Specialty Enzymes and Biochemicals' product portfolio consists of 160 unique enzyme solutions that address the needs of customers in 25 industries and over 30 countries. Our enzymes are all natural and non-GMO. They are a safe and effective alternative to the harsh chemicals often used in the manufacturing processes of many industries.

Industrial categories are:

- Alcohol
- Brewing
- Wine
- Food Processing
- Animals
- Baking
- Dairy
- Juice, Coffee and Tea
- Detergents
- Starch Processing
- Personal Care
- Pharmaceutical
- Textiles, Leather and Paper Pulp

Beverage Alcohol and Fuel Ethanol Industries

Enzymes are used in the alcohol processing industries to hydrolyze (break down) large starch molecules into simple sugars. Typically, the hydrolysis of starch to simple sugars is accomplished using a two-step, enzyme-catalyzed process called liquefaction and saccharifaction.

To save energy in alcohol production, starch can also be liquefied at lower temperatures using Specialty Enzymes' SEBamyl-BAL or fungal alpha-amylase SEBamyl-L. These alpha-amylases quickly reduce the viscosity of gelatinized starch. In distilleries with pre-saccharification systems, our glucoamylase blend SEBamyl-GL is used to convert liquefied-starch to fermentable sugars. SEBamyl retains excellent activity over a wide temperature range, making it suitable for use in simultaneous saccharification and fermentation distilleries.

Protease enzymes that breakdown proteins to their constituent amino acids are commonly added to whole-grain, alcohol fermentations to enhance yeast nutrition and alcohol yields. SEB-Neutral P increases free-amino nitrogen concentrations and the extraction of encapsulated starch when added to the fermentor. Non-starch, polysaccharide-hydrolyzing enzymes, such as CelluSEB-L and ViscoSEB-L, are used in cereal-grain fermentations to reduce mash viscosity, improve alcohol yields and reduce fouling of distillation columns and spent-beer evaporators.

Brewing

Enzymes have proved to be useful for the brewing industry in many areas of beer production. They can be added to the beer after its fermentation to induce faster maturation. Enzymes also work as filtration improvers, reducing the presence of viscous polysaccharides such as xylans and glucans. Enzymes are often used to remove carbohydrates in the production of light beer and to induce chill proofing. But most importantly, enzyme supplementation is simpler and less expensive than the malting process.

Beer brewing involves the production of alcohol by allowing yeast to act on plant materials such as barley, maize, sorghum, hops and rice. Yeast converts simple sugars into alcohol and carbon dioxide. However, the sugars found in the plant materials are most often complex polysaccharides that yeast is unable to convert. The traditional method for breaking down these complex polysaccharides are called malting. This is the process whereby barley, for example, is allowed to partially germinate, producing enzymes that break down the complex polysaccharides into simple sugars that the yeast can utilize. However, the process of malting can be expensive and often difficult to control. By adding enzymes to unmalted barley the complex polysaccharides can be broken down to simple sugars and reduce or eliminate the costly and complicated process of malting.

Wine

Many of the biochemical reactions involved in wine production are enzyme-catalyzed. They begin during the ripening and harvesting of grapes, and continue through alcoholic and malolactic fermentation, clarification, and ageing. Winemakers often supplement naturally occurring grape enzymes with commercial enzymes to increase production capacity of clear and stable wines with enhanced body, flavour and bouquet. When added to grapes or musts, our LiquiSEB products increase free-run juice volume and extraction of colour, fermentable sugars and flavour components, as well as reduce pressing and fermentation time.

These pectinase or pectinase containing hemicellulase products can increase free-run juice volume by 20 to 30 per cent and lower fermentation time by 30 to 50 per cent by reducing grape-pectin viscosity. Rapid clarification and well-separated lees have a positive effect on finished wine flavour, texture and colour. Our purified ClariSEB pectinases depectinize grape-musts during fermentation or young wines prior to fining and filtration. Grape musts and wines treated with ClariSEB are less viscous. They ferment, settle and mature more quickly. ClariSEB, a beta-glucanase containing pectinase, is also used to degrade Botrytis-glucan. Wines made from overripe grapes infected with Botrytis cinerea mold are often difficult to clarify and filter due to high concentrations of Botrytis-produced glucan polysaccharides. The use of ClariSEB can speed up clarification and filtration. Our SEB-Acid and SEB-Prolase acid proteases clarify and stabilize some wines by reducing or removing naturally occurring and yeast synthesized, heat-labile proteins.

Animal Feed and Pet Care

We have blends that help provide natural relief to dry or scaly hair coats, skin problems, digestive disorders, joint difficulties, immune disorders, excessive shedding, weight problems, allergies, bloating, lethargy, flatulence, and coprophagia. We also have a proprietary blend called Exclzyme®Pet that promotes tissue, muscle and wound healing, helps along recovery from soreness and inflammation and is helpful for pets who are suffering from arthritis and pain. Exclzyme®Pet is ideal for active or athletic pets.

Baking

Enzymes are rapidly becoming very important to the Baking Industry. Enzyme supplements are used in baking to make consistently high-quality products by enabling better dough handling, providing anti-staling properties, and allowing control over crumb texture and colour, taste, moisture, and volume. Enzymes are now replacing chemical supplements that were used in the baking industry to account for these discrepancies. For instance, our enzyme called SEBake-GO can replace chemical oxidants, such as bromates, used to strengthen gluten.

Dairy

Specialty Enzymes' SEB-RMP, a microbial rennet used in cheese production, is an effective substitute for calf/cow, goat, sheep or swine-derived rennets. When milk is treated with milk-clotting proteases or microbial proteases, it causes casein to precipitate in the form of curd. The curd is separated from other milk ingredients (whey) and further processed to produce soft and hard cheeses. SEB-RMP is a superior option to animal derived rennets. Microbial lipases and proteases can be cost-effectively added to fresh milk or cheese curd to produce products with enhanced dairy and cheese flavours. Specialty Enzymes' fungal lipases SEB-ATlipase and SEB-CClipase are used to enhance the dairy-flavour in coffee whiteners, margarine and snack foods. These enzyme blends also accelerate the ripening of naturally aged cheeses and produce intensely-flavoured, enzyme-modified cheeses (EMC) used as a flavour in snack foods, soups, frozen dinners and cheese-sauces. Protease enzymes also play an important role cheese ripening and flavour development. The production of strong cheese-flavours is enhanced when fungal lipases and Specialty Enzymes' SEB-Npro, a neutral protease, or SEB-Pro, a fungal protease, are added to fresh cheese curd. It is possible develop a variety of intense cheese flavour-profiles in a few hours or days when using these protease and lipase enzymes together.

Specialty Enzymes offers SEB-Alkaline, SEB-Prolase, and SEBLipo, which are to be added with mild detergents and pH adjusting agents. These solutions will help remove milk protein and fat deposits from the filter, increasing filtration rates without using the harsh chemicals that add to waste effluent treatment costs.

Food Processing

SEB-Brolase and SEB-Prolase rapidly hydrolyze both animal and plant proteins to reduce or eliminate gel formation on heating. They are used to hydrolyze soy, pea, yeast, maize and wheat-flour proteins to produce non-bitter, protein hydrolysates. Pre-processed or separated plant and milk proteins such as soya isolates and concentrates, casein and whey can be easily hydrolyzed to produce non-bitter protein hydrolysates suitable for use as active ingredients in dietary supplements, functional beverages, fermented foods, baked goods and other foods.

SEB-Acid, SEB-Alkaline and SEB-Neutral are different types of protease enzymes used in food processing applications to break down and increase the solubility, dispersibility and digestibility of proteins. They also hydrolyze vegetable and animal proteins to peptides, polypeptides, and small amounts of amino acids. SEBzyme-XL and CelluSEB is used for extraction, for liquefaction of plant materials and for downstream processing in the food, starch, and alcohol and brewing industries. In the food industry, these products are used to improve the processing and extraction of wheat and corn gluten, soybean protein, cereal starch, seed and nut oils, flavours and colours, coffee, tea and other botanical extracts.

Fruit and Vegetable Juice

Specialty Enzymes products, LiquiSEB and ExtractSEB, are pectinases that contain hemicellulase enzyme activities. These products not only increase juice yields, but also increase the colour and health-promoting antioxidants in fruit and vegetable juices extracted by pressing or decanter centrifuge. By reducing fruit and vegetable mash viscosity and improving solid/liquid separation, they increase colour extraction and juice volume by as much as 15 per cent.

ClariSEB and SEBamyl enzyme solutions speed up filtration and prevent storage or post-packaging haze formation by depectinizing and reducing starch in raw juices. Pectin and starch must be removed from freshly extracted juices prior to filtration, fining and concentration. Unripe fruits and vegetables can contain as much as 15 per cent starch. ClariSEB pectinase and SEBamyl amylase can reduce starch and pectin in raw fruits and juices, thus achieving clear and stable juices and juice concentrates.

Coffee and Tea

Bioclean TC has a blend of natural enzymes that gives manufacturing areas a 100 per cent safe-detergent and alkali-free cleansing. Bioclean TC is dissolved in water and sprayed in areas that come in contact with leaf material. The blend helps degrade biological material, which is then easily dislodged in cold water washing. Bioclean TC prevents bacterial growth by not providing any food to sustain the bacteria. And because it is 100 per cent natural, the blend doesn't impart any off-flavour or off-taste to the tea. Specialty Enzymes, a leading solution provider to the human healthcare, animal nutrition, and natural product processing industries, offers SPEEDOX. When added to the wash water or effluent stream in the coffee production process, SPEEDOX rapidly reduces organic loads in the wastewater that cause environmental pollution. SPEEDOX can reduce the pollution causing entities in coffee plantation wastewater. SPEEDOX is a 100% natural, high quality and eco-friendly product. It improves the clarity of water so the water can be used for irrigation. SPEEDOX also eliminates the foul odour near the effluent treatment tanks.

INDUSTRIAL BIOTECHNOLOGY

At present, a third wave of biotechnology-industrial biotechnology-is strongly developing. Industrial biotechnology (also referred to as white biotechnology) uses biological systems for the production of useful chemical entities. This technology is mainly based on biocatalysis and fermentation technology in combination with recent breakthroughs in the forefront of molecular genetics and metabolic engineering. This new technology has developed into a main contributor to the so-called green chemistry, in which renewable resources such as sugars or vegetable oils are converted into a wide variety of chemical substances such as fine and bulk chemicals, pharmaceuticals, bio-colorants, solvents, bio-plastics, vitamins, food additives, bio-pesticides and bio-fuels such as bio-ethanol and bio-diesel. The application of industrial biotechnology offers significant ecological advantages. Agricultural crops are used starting raw materials, instead of using fossil resources such as crude oil and gas. This technology consequently has a beneficial effect on greenhouse gas emissions and at the same time supports the agricultural sector producing these raw materials. Industrial biotechnology frequently shows significant performance benefits compared to conventional chemical technology.

Strategic Actions:

- Focus in industrial biotechnology will be on reducing chemical and toxic load in our effluent streams, developing non-fossil fuels that are eco-friendly and developing green technologies in Industrial processing.
- Encourage public-private partnership to promote investment in this sector.
- Promotion of industrial biotechnology in strategic areas of manufacturing and developing green technologies.

PREVENTIVE AND THERAPEUTIC MEDICAL BIOTECHNOLOGY

A healthy population is essential for economic development. Important contributors to the total disease burden are infections like HIV-AIDS, tuberculosis, malaria, respiratory infections and chronic diseases affecting the heart and blood vessels, neuro-psychiatric disorders, diabetes and cancer. It is important to synchronize the technology and products with the local needs of the health system and to facilitate technology diffusion into health practice. Increasing knowledge about pathogen genomes and subtypes, host responses to infectious challenges, molecular determinants of virulence and protective immunity and novel understanding mechanisms underlying escaped immunity and ways to develop novel immunogens will guide development of vaccines against infectious diseases. Translational research and ability to rapidly evaluate multiple candidates in clinical trials can help accelerate the pace of vaccine development.

New directions in manufacturing and delivery are emerging. Major opportunities to control costs are the more efficient processes for manufacturing of new pharmaceuticals, more efficient systems for production of therapeutic proteins and biomaterials and development of drug delivery systems that release drugs at a target site. A shift from parenteral to oral or transcutaneous administration of drugs and vaccine holds the promise of simplifying delivery in health systems. Medical biotechnology offers a significant possibility for Indian industry to establish a strong pharmacy sector, a growing number of small and medium biotechnology companies, a large network of universities, research institutes, and medical schools and low cost of product evaluation. The medical biotechnology sector annually contributes over 2/3rd of the biotechnology industry turnover. The Indian vaccine industry has highlighted India's potential by emerging as an important source of low cost vaccine for the entire developing world.

Further, economic opportunities through contract research and manufacturing through global partnerships are large if supported by enabling government policies and incentives. The policy goal is to accord high priority to basic and applied research, to strengthen capacity in pre-clinical and clinical product evaluation technologies relevant to all aspects of health and medical care-predictive, preventive, therapeutic and restorative will be supported. Innovation will be supported through new granting mechanisms to support interdisciplinary networks and public private partnerships.

Strategic Actions

(i) Research emphasis

- Basic and applied research would be supported in molecular and cellular biology, genomics, proteomics, system biology, stem cell biology, RNA interference, host response and new platform technologies.
- Pathogenesis of major diseases and molecular mechanics of disease transmission would be investigated
- Product development will be focused on vaccines, diagnostics, new therapies based on cell and tissue replacement, therapeutic antibodies, herbal medicine, plant based medicine, nucleic acids, therapeutics, drug and vaccine delivery systems, new anti microbial agents
- Research to improve production and manufacturing process and local production of biological reagents for development of diagnostics will be supported.

(ii) Improvements in infrastructure and networks

- A centre for translational research will be established. This new institute will be interdisciplinary and will deal with technology policy for public health, molecular pathogenesis of disease, technology development, scale up, product evaluation and technology diffusion into programmes. Centre will be unique in having a pool of scientists, physicians, engineers, and public health persons working on public health grand challenges. This institute will work through public-private partnerships and be a training centre for product development, IPR and regulation.
- A mission mode programme will be initiated in biomaterial and medical device area as an integrated effort by the

Department of Science & Technology and Department of Biotechnology. The goal is to promote R&D and industrial activity.

- Two centres of molecular medicine will be supported within medical school system closely interacting with basic science institutes.
- A virtual network of stem cell centres will be established, using a city cluster approach to network scientists and clinicians. Two core stem cell research centres will be established together with several network sub-clusters. An umbilical cord stem cell bank will be established.
- Mechanism based screening of herbal drugs known in traditional Indian systems would be carried out so as to get value added therapeutics products quickly
- An inter agency task force of ICMR, Department of Biotechnology, and DST will be established to suggest strategies for strengthening medical school based research. Capacity related to translational biology, clinical trials, molecular epidemiology and product development would be strengthened. Integrated MD-PhD programmes will be supported.

(iii) Streamline guidelines and procedures for the approval of recombinant pharmaceutical products. Currently there are multiple regulators, multiple ministries, lack of coordination between these regulators, Over-lapping and duplication of responsibilities of these regulators, lack of a linear progression in the approval process and committees working outside their area of expertise. The Mashelkar Committee has drawn up a new procedural framework for Biopharmaceuticals, which has streamlined the regulatory process:

- IBSC will monitor all development work (upto 20 Litres) and recommend to RCGM for Animal Toxicity Tests (ATT) & Scale up.
- RCGM will evaluate the recombinant technology & grant permission for scale up-R & D, review and approve for preclinical animal toxicity tests and evaluate ATT data & recommend to DCGI for Human Clinical Trial (HCT).
- DCGI will permit Human Clinical Trials, review Human Clinical Trial Data, grant permission for Manufacture and Marketing the product and inspect the facility where product is manufactured.
- GEAC will review the manufacturing process to ensure that the LMO (living modified organism) is "inactivated" during

the process and send its recommendations to the Drugs Controller General of India within the specified time. The GEAC would confine its approval role to LMOs and Category 3 and 4 microorganisms.

AGRICULTURE COMPETITIVE

Biotechnology is necessary to maintain our agriculture competitive and remunerative and to achieve nutrition security in the face of major challenges such as declining per capita availability of arable land; lower productivity of crops, livestock and fisheries, heavy production losses due to biotic (insects pests, weeds) and abiotic (salinity, drought, alkalinity) stresses; heavy post-harvest crop damage and declining availability of water as an agricultural input. Investment in agricultural related biotechnology has resulted in significantly enhanced R&D capability and institutional building over the years. However, progress has been rather slow in converting the research leads into usable products.

Uncertainties regarding IPR management and regulatory requirements, poor understanding of risk assessment and lack of effective management and commercialization strategies have been significant impediments. India owns very few genes of applied value. The majority of the genes under use-about 40-are currently held by MNCs and have been received under material transfer agreements for R&D purpose without clarity on the potential for commercialization.

The spectrum of biotechnology application in agriculture is very wide and includes generation of improved crops, animals, plants of agro forestry importance; microbes; use of molecular markers to tag genes of interest; accelerating of breeding through marker-assisted selection; fingerprinting of cultivars, land raises, germplasm stocks; DNA based diagnostics for pests/ pathogens of crops, farm animals and fish; assessment and monitoring of bio diversity; in vitro mass multiplication of elite planting material; embryo transfer technology for animal breeding; food and feed biotechnology. Plants and animals are being used for the production of therapeutically or industrially useful products, the emphasis being on improving efficiency and lowering the cost of production. However, emphasis should not be on edible vaccines for which use in real life condition is difficult. Nutrition and balanced diet are emerging to be important health promotional strategies. Biotechnology has a critical role in developing and processing value added products of enhanced nutritive quality and providing tools for ensuring and monitoring food quality and safety.

It has been estimated that if Biofertilizers were used to substitute only 25% of chemical fertilizers on just 50% of India's crops the potential would be 2,35,000 MT. Today about 13,000 MT of Biofertilizers are used-only 0.36% of the total fertilizer use. The projected production target by 2011 is roughly around 50,000 MT. Biopesticides have fared slightly better with 2.5% share of the total pesticide market of 2700 crores and an annual growth rate of 10-15 %. In spite of the obvious advantages, several constraints have limited their wider usage such as products of inconsistent quality, short shelf life, sensitivity to drought, temperature, and agronomic conditions. From a research perspective the spectrum of organisms studied has been rather narrow and testing has been on limited scale and restricted mainly to agronomic parameters.

Environmental factors such as survival in the Rhizosphere/phyllosphere and competition of native microbes have not received sufficient attention. Moreover, results on crops are slow to show. Unless there is a policy initiative at the centre and the state to actively promote Biofertilizers and biopesticides at a faster pace, there is unlikely to be a quantum jump in their consumption. A taskforce headed by Dr MS Swaminthan under the Ministry of Agriculture has prepared a detailed framework on the application of biotechnology in agriculture.

The report rightly lays emphasis on the judicious use of biotechnologies for the economic well being of farm families, food security of the nation, health security of the consumer, protection of the environment, and security of national and international trade in farm commodities.

CROPS AND TRAITS OF PRIORITIES

Priorities for crops and traits should be set after conducting a need assessment exercise in various farming zones. However, an indicative list has been suggested by MS Swaminathan Task Force.

Priority target traits in crop plants would be yield increase, pest and disease resistance, abiotic stress tolerance, enhanced quality, and shelf life, engineering male sterility and development of apomixis. Crops of priority should be rice, wheat, maize, sorghum, pigeon pea, chickpea, moong bean, groundnut, mustard, soybean, cotton, sugarcane, potato, tomato, cole crops, banana, papayas and citrus.

In priority crops equal emphasis should be given to GM hybrids and new varieties. The varieties in contrast to hybrids, are preferred by small farmers as they can use their own farm saved seeds for at least three or four years. In case of hybrids, research on the introduction of genetic factors for apomixis would be supported so that resource-poor farmers can derive benefits from hybrid vigour without having to buy expensive seeds every cropping season.Priority target traits in livestock would be enhanced fertility and reproductive performance, improved quality, resistance to diseases for reduced drug use, production of therapeutically useful products and quality feed. Livestock of priority would be buffalo, cattle, sheep and goat. Emphasis would be given to animal healthcare, nutrition, development of transgenics and genomics. It is proposed to set up an autonomous institution for animal biotechnology.

Application of biotechnology would be crucial in disease resistance, enhanced productivity, fertility and reproductive growth, use of aquatic species as bioreactors for production of industrial products, value added products from sea weeds and other marine taxa and biosensors for pollution monitoring. Species of priority in fisheries would be carps, tiger shrimps and fresh water prawns. It is proposed to set up under the auspices of DBT and autonomous centre for marine biotechnology

R&D would be focused on: development of biotechnology tools for evaluating food safety, development of rapid diagnostic kits for detection of various food borne pathogens; development of analogical methods for detection of genetically modified foods and products derived there from; development of nutraceuticals/health food supplements/functional foods for holistic health; development of pre-cooked, ready-to-eat, nutritionally fortified food for school going children; development of suitable pro-biotics for therapeutic purposes and development of bio food additives. It is proposed to set up (under the auspices of Department of Biotechnology) an autonomous institute for nutritional biology and food biotechnology.

Priorities would include screening of elite strains of micros-organisms and/or productions of super-strains, better understanding of the dynamics of symbiotic nitrogen fixation, process optimization for fermentor-based technologies, improved shelf life, better quality standards, setting up accredited quality control laboratories and standardization of GMP guidelines. Integrated nutrient management system would be further strengthened.

6

Energy Nutrients

The need for energy nutrients is high in athletes, and a great deal of science focuses on optimal distribution of the energy substrates to support exercise of different intensities and durations. Studies have demonstrated clearly that dependence on carbohydrate increases with greater exercise intensity, and many studies provide valuable guidelines on how best to deliver carbohydrate to optimize glycogen stores, how best to deliver carbohydrate during training and competition, and how energy substrates contribute to muscle recovery.

The contribution of protein to muscle function and recovery is much better understood now than in the recent past, and there is some new science on the relationship between mental and muscle function that is mediated by carbohydrate, protein, and fat. The recent popularity of higher-protein, higher-fat, and lower-carbohydrate diets has serious implications for athletic performance. It is critical that athletes and coaches understand how to assess the appropriate energy intake and energy substrate distribution to optimize both mental and muscle function. This chapter presents the essential elements of carbohydrate, protein, and fat metabolism in exercise, along with a critical scientific view on how these substrates contribute to optimal athletic performance.

CARBOHYDRATES

There are different types of carbohydrate, and each type is treated differently by our bodies. For instance, glucose and bran are both carbohydrates, but they are on different ends of the energy spectrum. Glucose enters the bloodstream quickly and initiates a fast and high insulin response, while the energy in bran never makes it into the bloodstream because of its indigestibility and tends to mediate the insulin response by slowing the rate at which other energy sources enter the bloodstream. These within-carbohydrate differences mandate that athletes should carefully consider the type of carbohydrate that might be best under different circumstances. Glucose is the main source of fuel for muscular activity, and the higher the exercise intensity, the greater the reliance on glucose as a fuel. When glucose runs out, the athlete stops performing. Therefore, understanding how to keep glucose from becoming depleted should become a major focus of an athlete's nutrition practices. Sustaining carbohydrate sufficiency is problematic because, unlike either protein or fat, humans have a limited storage capacity for carbohydrate. Carbohydrate adequacy becomes even more critical at higher levels of exercise intensity because there is a greater reliance on carbohydrate as a source of muscular fuel. Despite years of research confirming the importance of maintaining carbohydrate availability for sustaining muscular endurance and mental function, many athletes still believe protein is the critical substrate for achieving athletic success. Although all substrates are important, delivering the right amounts of carbohydrate at the right time optimizes the limited carbohydrate stores, ensures better carbohydrate delivery to the brain, and improves endurance performance. By comparison, the common focus on excess protein consumption does little to enhance performance or a sense of well-being.

CARBOHYDRATE METABOLISM

Humans can store approximately 350 grams (1,400 kilocalories) in the form of muscle glycogen, an additional 90 grams (360 kilocalories) in the liver, and a small amount of circulating glucose in the blood (~5 grams, or about 20 kilocalories). The larger the muscle mass, the greater the potential glycogen storage but also the greater the potential need. We have systems for maintaining blood glucose within a relatively narrow range (70 to 110 milligrams per deciliter) by recruiting insulin and glucagon. Insulin and glucagon are pancreatic hormones that work synergistically to control blood glucose. Excess production of insulin can result in hypoglycemia (low blood sugar), with a resultant excess production of fat; inadequate insulin production results in hyperglycemia (high blood sugar) and diabetes.

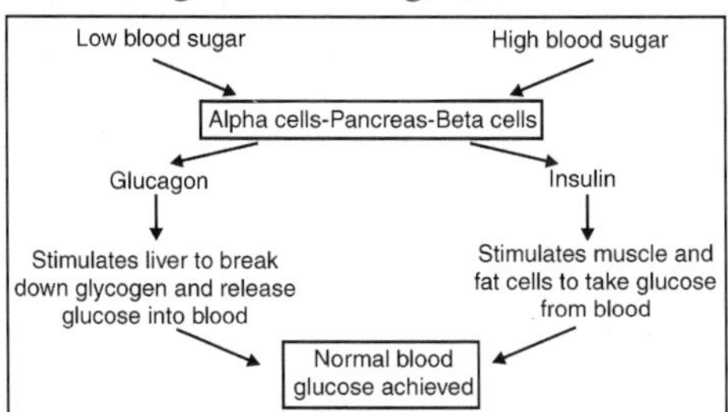

Fig. The Impact of the Pancreas on Normalizing the Blood Glucose Levle.

Insulin is secreted by the beta cells of the pancreas, whereas glucagon is secreted by the alpha cells of the pancreas. The stimulus for insulin secretion is high blood glucose (the higher the glucose, the higher the insulin response), but a small amount of insulin is constantly being secreted by the pancreas even when blood glucose is in the normal range, causing a steady flow of glucose to the cells of the brain and muscles. Insulin lowers blood glucose by affecting the cell membranes of muscle and fat cells, thereby allowing glucose from the blood to enter the cell. This action causes a transfer from blood glucose to cell glucose and explains the blood-glucose-lowering effect of insulin; it also enables cells to receive a needed source of energy.

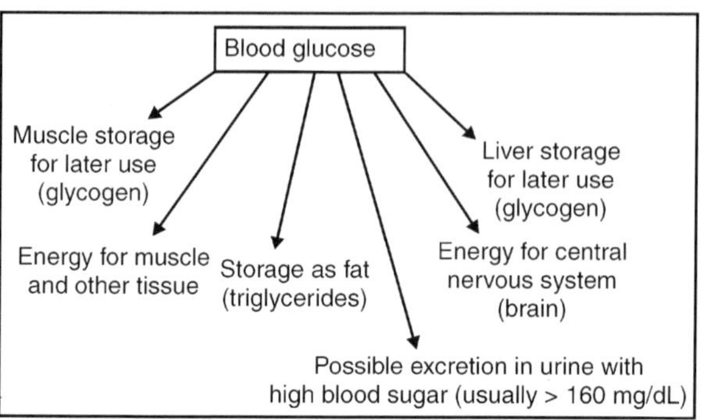

Fig. Possible Pathways Taken by Blood Glucose.

With low blood glucose, as occurs between meals and during exercise, glucagon is secreted. Lower levels of blood glucose result in greater glucagon production. Glucagon causes catabolism of liver glycogen, which results in the release of some of its glucose molecules into the blood. Glucagon may also stimulate gluconeogenesis (the manufacture of glucose from nonglucose substances). The amino acid alanine, for instance, is derived from protein and is converted to glucose by the liver.

About 60 percent of the glucose released by the liver to sustain blood sugar is from liver glycogen stores, but the remainder is from glucose synthesized from lactate, pyruvate, glycerol, and amino acids. The rate of liver glucose infused into the blood during exercise is a function of exercise intensity, with higher-intensity exercise causing a faster rate of liver glucose release. The combination of lower blood insulin and higher epinephrine and glucagon during long-duration activity stimulates liver glucose release.

Besides insulin and glucagon, two other hormones also influence blood glucose. Epinephrine (adrenaline) is a stress hormone that initiates an extremely rapid breakdown of liver glycogen to quickly increase blood glucose levels. Cortisol, which is secreted from the adrenal gland, is also a stress hormone that promotes protein catabolism. This protein breakdown makes certain glucogenic amino acids available for gluconeogenesis, ultimately resulting in an increase in blood glucose. Both epinephrine and cortisol are released as a result of exercise-related stress, and both can be mediated through maintenance of blood glucose. Controlling epinephrine production helps preserve liver glycogen, and controlling cortisol helps preserve muscle protein. This is a strong argument for consuming carbohydrate during exercise.

The glucose circulating in the blood is derived mainly from dietary carbohydrate, with starch constituting the major source. Complex carbohydrates (starches) are digested into monosaccharides (glucose, fructose, and galactose) for absorption into the blood. In some cases, individuals have inadequate lactase to break milk sugar (lactose) into its component monosaccharides (glucose and galactose), causing the lactose to go indigested in the gut. Referred to as *lactose intolerance,* this leads to bloating, abdominal pain, diarrhea, and dehydration.

Excess glucose in the liver and muscle is stored as glycogen, but only up to the glycogen saturation point. The liver has a maximum glycogen storage capacity of approximately 87 to 100 grams (348 to 400 kilocalories), while the muscles can store approximately 350 grams (1,400 kilocalories), or more in larger individuals. Providing additional glucose to cells when glycogen stores are saturated leads to the excess being stored as fat (in both muscle and fat cells). The liver glycogen is primarily responsible for stabilizing blood glucose, while the muscle glycogen is mainly responsible for providing an energy source to working muscles that can be used both aerobically and anaerobically. Blood sugar is not easily maintained when liver glycogen is depleted, even when muscle glycogen stores are full.

Blood glucose is the primary fuel source for the central nervous system. Low blood sugar results in depressed central nervous system activity, coupled with increased irritability and a lower capacity to concentrate. For athletes, low blood sugar may be related to mental fatigue, which is related to muscle fatigue. Because liver glycogen and blood glucose stores are easily depleted during even short-duration activities, the intake of carbohydrate during activity is a critical factor in maintaining mental function and, ultimately, muscle function.

GLYCOLYSIS

ATP (adenosine triphosphate) is the high-energy compound for cells. We have a limited storage of immediately available ATP, so it must be generated quickly during exercise. The higher the exercise intensity, the faster the ATP must be regenerated. In steady-state, low-intensity activity, ATP can be adequately produced aerobically from the oxidation of carbohydrate and fat. However, as exercise intensity increases, athletes need a level of ATP production that cannot be fully supplied aerobically.

CARBOHYDRATES AND PHYSICAL ACTIVITY

Because physical activity dramatically increases the rate of energy expenditure, athletes must strategize in order to best supply the needed energy to achieve success. It is critical for athletes to obtain sufficient total energy intake to support total energy requirements, including those for normal tissue maintenance, growth (in children and adolescents), tissue repair, and the energy requirements of the activity itself. It is impossible to discuss the ideal distribution of energy substrates without first conceptualizing how to best meet total energy needs. Although this may seem a logical and simple act, virtually all surveys have found that athletes fail to consume sufficient energy to fully satisfy their needs. This is much like planning to take a Ferrari on a 100-mile trip, and you appropriately put high-octane gas in the fuel tank-but only enough of it to go 80 miles. The Ferrari just won't get there, and poorly fueled athletes will also have difficulties doing what's required to be optimally competitive. Once a strategy has been established for obtaining sufficient energy, then athletes can reasonably consider how to parse the energy into the optimal distribution of energy substrates: carbohydrate, protein, and fat. It is generally accepted that athletes should consume sufficient carbohydrate to meet the majority of their exercise-related energy needs, plus enough carbohydrate to restore muscle glycogen stores between exercise sessions. Ideally, athletes should consume complex carbohydrates when possible but should consume simple carbohydrates during and immediately after exercise. Other energy substrates (protein and fat) should also be consumed to fulfill total nutrient requirements, but carbohydrate should remain the predominant energy source. It is difficult for athletes to take in sufficient energy and carbohydrate unless there is a well- established plan to do so. Athletes would do well to remember that training alone, without a sound nutrition plan to support the training, will be self-limiting.

FAT (LIPIDS)

Despite some recent literature wrongly espousing the benefits of high fat intakes (i.e., intakes of 30 percent or more of total calories from fat), fat is a highly concentrated fuel that does nothing to improve athletic performance, body composition, or weight when taken in excess. The adult AMDR for total fat intake is 20 to 35 percent of total calories, and there is no scientific information suggesting that more than 25 percent of total calories from fat is generally better for athletes.

However, for athletes who may have difficulty sustaining weight because of a massive energy expenditure (such as cross-country skiers) or who must sustain high weights (such as linemen on football teams), higher fat intakes (up to the AMDR limit of 35 percent) may be necessary. Few Americans consume less than 35 percent of total calories from fat, so consumption of less fat is not easy, and unless steps are taken to provide sufficient energy from other substrates (mainly from more complex carbohydrates) to replace the eliminated fat, athletes may place themselves in an energy-deficit state that is, in itself, a detriment to performance. Therefore, while a reduction in fat intake is generally useful, a conscious effort should be made to provide enough total energy when fat intake is reduced.

Since fat is more than twice as concentrated in calories than either protein or carbohydrate (9 calories per gram versus 4 calories per gram), more than twice as much food must be consumed to make up the difference in reduced fat. Cholesterol, oils, butter, and margarine are all fats, or lipids, but each has slightly different characteristics. The one common attribute shared by lipids is that they are soluble in organic solvents but not soluble in water. (Anyone who has tried to mix Italian dressing knows this to be true.

The oil in the dressing eventually rises to the top, no matter how hard the bottle is shaken.) The term *fat* is usually applied to lipids that are solid at room temperature, and the term *oil* is applied to lipids that are liquid at room temperature. The most commonly consumed form of lipid is triglyceride, which consists of three fatty acids and one glycerol molecule (thus the name tri glyceride). Despite the numerous forms of lipids, we can obtain them all from the food supply, and we are also capable of making many types of lipids by combining carbon units from other substances. Nearly every cell in the body has the capacity to make cholesterol, which is why a person can have a high blood cholesterol level even when on a low-cholesterol diet. We can also manufacture phospholipids, triglycerides, and oils. In fact, it is this ability to effectively manufacture different types of lipids that limits the necessity to consume large amounts of lipids.

FAT FUNCTIONS

A certain amount of fat, between 20 and 35 percent of total consumed calories, is necessary to ensure a sufficient energy and nutrient intake. The fat-soluble vitamins-vitamins A,D,E, and K-must be delivered in a fat package. The essential fatty acids, which are needed for specific body functions but that we are incapable of synthesizing, must also be consumed. Some dietary fat is also needed to give us a feeling of satiety during the meal, creating the important physiological signal that it is time to stop eating. Dietary fats have a longer gastric emptying time than do carbohydrates, which contributes to the feeling of satiety. Of course, fat also makes foods taste good.

LIPID STRUCTURE

Lipids have different levels of saturation, a term that refers to the number of double bonds in the carbon chain. Fatty acids with no double bonds are saturated, those with one double bond are monounsaturated, and those with more than one double bond are polyunsaturated. Single bonds are stronger and less chemically reactive than double bonds, so the greater the number of double bonds, the greater the opportunity for the fatty acid to react with its chemical environment. It is this differential reactive capacity that makes the number of double bonds an important factor in human nutrition. Saturated fatty acids are most prevalent in fats of animal origin, palm kernel oil, and coconut oil. Monounsaturated fats are highest in olive oil and canola oil but are also present in fats of animal origin. Polyunsaturated fats are highest in vegetable oils (with the exception of olive oil, which is more than 75 percent monounsaturated). In the context of a fat intake that does not exceed 35 percent of total calories, mono- and polyunsaturated fats should make up the majority of the fats consumed. Saturated fats are associated with higher cholesterol levels, so they should be minimized when possible. This is most easily achieved by reducing the consumption of animal fats, chocolate candies (often high in saturated tropical oils), fried foods, and high-fat dairy products.

TRIGLYCERIDES

The majority of consumed lipids are triglycerides, which contain three fatty acids and a glycerol molecule. Fat is stored in the form of triglycerides, which we manufacture when excess energy is consumed. We store triglycerides in adipose tissues (groups of fat cells) and inside muscle cells (intramuscular tri glyceride), both of which are available as an energy source when needed. When fat is burned as a source of energy, the stored triglycerides are taken out of storage, and each molecule is cleaved into its component fatty acids and glycerol molecule. Each fatty acid can then be broken apart (two carbon units at a time) and thrown into the cellular furnaces for the creation of ATP to form heat and provide the energy for muscular work. This process is referred to as the beta-oxidative metabolic pathway because burning fat, besides requiring some carbohydrate for its complete oxidation, also requires oxygen.

Glycerol is a unique lipid that is burned like a carbohydrate rather than a fat and is also an effective humectant (it holds water). Some long-endurance athletes find that adding glycerol to water helps them retain more water (i.e., to superhydrate) than if they consumed water alone. In an extremely hot and humid environment, water loss is likely to be higher than the athlete's fluid-replacement capacity, so beginning a competition in a superhydrated state may provide some advantages. In studies of tennis and Olympic distance triathlon, athletes consuming glycerol in water prior to competition experienced some protective hyperhydration benefits when exercising in high heat. However, being superhydrated is likely to impart a degree of discomfort that requires some adaptation. Athletes who consume fluids containing glycerol often describe the feeling of holding extra body water as making them feel "like a water bag," or "heavy," or "stiff. " Still, they maintain that how they feel at the end of a race is more important than how they feel at the beginning, so adding glycerol to precompetition fluids has become a standard protocol for some athletes.

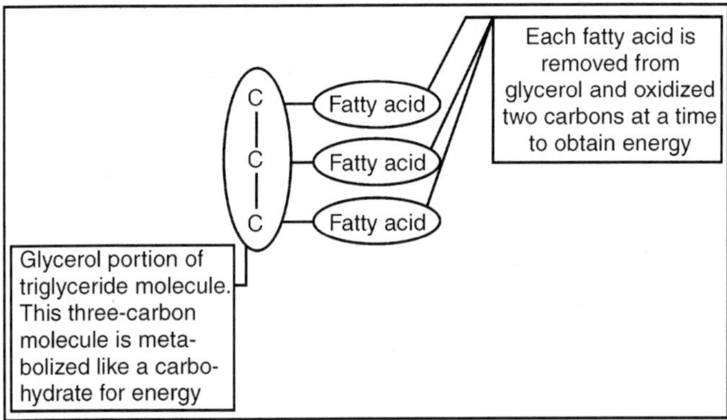

Fig. Triglyceride Structure.

ESSENTIAL FATTY ACIDS

Linoleic (omega-6) and linolenic (omega-3) fatty acids are the essential fatty acids; although they are needed for metabolic processes, we are incapable of synthesizing them. The "omega-6" classification means that these fatty acids, which are polyunsaturated, have the last double bond six carbons from the end of the carbon chain. The "omega-3" classification means that these fatty acids have the last double bond three carbons from the end of the carbon chain. Linoleic acid is an essential part of lipid membranes and is required for normal skin health. Linolenic acid is necessary for neural function and growth.

The AMDR for omega-6 and omega-3 fatty acids is 5 to 10 and .6 to 1.2 grams per day, respectively. Both fatty acids are easily obtained from vegetable oils (corn, safflower, canola, and so on) and the oils of cold-water fish. Fish liver oils that are high in omega-3 acids have recently received much attention. These oils have been shown to reduce the ability of red blood cells to congregate, thereby decreasing the chance for an unwanted blood clot to form. This reduces the risk of a heart attack, which is most commonly caused by a clot formation in one of the major heart arteries. The oils from cold-water fish are the main source of the omega-3 fatty acids eicosapentanoic acid (EPA) and docosohexanoic acid (DHA). Even a once-weekly consumption of cold-water fish (salmon, albacore tuna, Atlantic herring) is sufficient to significantly reduce the risk of heart attack and stroke.

Despite these findings, excessive intake of these fish oils may cause problems, including an increase in cellular oxidative damage. The best rule of thumb is to make fish consumption a regular part of the weekly diet so that supplemental intake of omega-3 fatty acids is unnecessary

FAT REQUIREMENTS

From an exercise standpoint, there is little reason to believe that increasing fat consumption results in improving athletic performance, unless the increase in fat intake is the only reasonable means for the athlete to obtain sufficient energy. For the athlete who needs more than 4,000 calories each day to meet the combined demands of growth, exercise, and maintenance, moderate increases in dietary fat (preferably from plant and fish sources) may be needed. Since fat is a more concentrated form of energy than either carbohydrate or protein, more energy can be consumed in a smaller food package if the foods contain more fat. If an athlete tries to restrict fat completely, so much food needs to be consumed that it may be impossible to schedule enough meals or enough time during meals to consume the needed energy, leading to an inadequate energy consumption.

LIPIDS AND PHYSICAL ACTIVITY

Even the leanest, healthiest athletes have a substantial energy pool of stored lipids. The average storage in adipose tissue ranges between 50,000 and 100,000 calories, or enough energy to walk or run 500 to 1,000 miles (800 to 1,600 kilometers) without a refueling stop. In addition, athletes store approximately 2,000 to 3,000 calories of lipids inside the muscle tissue. These lipids, which are stored in the form of triglycerides, are available as a fuel under the proper conditions of oxygen availability. Maximal fat oxidation occurs at 60 to 65 percent VO_2 max, but at higher levels of VO_2max there is insufficient oxygen to derive the majority of energy used from fat catabolism.

Triglycerides stored in adipose tissue are cleaved into their component molecules of glycerol and fatty acids and transported to the blood plasma. The glycerol is available to all tissues for energy metabolism, and the free fatty acids are transported to working muscles where they are oxidized for energy. The triglycerides stored in working muscles are cleaved into glycerol and fatty acids, and the fatty acids can be oxidized for energy where they are resident. The glycerol can also be burned for energy in the working muscle or can be transported to the blood plasma as a source of energy for other tissues.

The lower the exercise intensity, the greater the proportion of fat burned to satisfy energy needs. As exercise intensity increases, the proportion of fat burned decreases and the proportion of carbohydrate burned increases. It is this basic reality that is behind why so many people do low-intensity activity to burn fat and lower body fat levels. However, the proportion of fat burned should not be confused by the total amount of fat burned at different intensities of physical activity. As exercise intensity increases, the total number of calories burned per unit of time also increases. Although there may be a decrease in the proportion of fat burned to satisfy total energy needs in higher-intensity activity, the total volume of fat burned is greater because the total energy requirement is higher. The take-away message from this metabolic reality is that athletes interested in lowering body fat should exercise at least as high as 65 percent of VO_2max for the duration of their workouts to optimize the total mass of fat that is burned. Exercising at lower intensities burns a greater proportion of fat but less total fat than exercising at higher intensities.

CARBOHYDRATE UTILIZATION DURING EXERCISE

Low carbohydrate levels result in exercise fatigue. Since carbohydrate stores (or, *glycogen stores*) are limited (~350 kilocalories of glycogen in the liver; ~1,400 kilocalories of glycogen in muscles), athletes should consider how to initiate exercise with glycogen stores full and should establish a routine that keeps glycogen stores from running low. Even if muscle glycogen stores are adequate, low liver glycogen stores will result in hypoglycemia and mental fatigue, and mental fatigue leads to muscle fatigue.

The higher the exercise intensity, the greater the reliance athletes have on carbohydrate as an energy substrate. However, even low-intensity (i.e., mainly aerobic) exercise that derives most of its fuel from fat still requires some level of carbohydrate for the complete combustion of fat and to maintain blood glucose. Therefore, all modes of physical activity have some degree of carbohydrate dependence.

Several factors influence the proportionate contribution of carbohydrate to total fuel requirements during exercise.

Factors that increase the reliance on carbohydrate include the following:

- High-intensity activity
- Long-duration activity
- Exercise in hot and cold temperature extremes
- Exercise at high altitude
- Age (higher in young boys than in men)

Factors that decrease the relative energy expenditure from carbohydrate include the following:

- Endurance training
- Good conditioning
- Temperature adaptation
- Gender

There is a common misconception that low-intensity activity (up to ~65 percent.VO_2max) is the most efficient means of fat loss. In fact, many popular exercise programs have been organized around the idea that a greater amount of fat is most efficiently "Burned" with low-intensity aerobic exercise. However, the proportion of fat burned should not be confused with the volume of fat burned. While you're sitting there reading this sentence, you are very likely deriving the vast majority of your energy requirements from fat.

However, the total volume of fat being burned is extremely low. (If this were not so, sitting in front of a TV would be a terrific means of initiating a fat-loss program.) When exercise intensity is increased, the proportion of energy derived from fat is decreased, and the proportion of energy derived from carbohydrate is increased, but some level of fat is always being burned. The total caloric requirement per unit of time is much greater in high-intensity activity than in low-intensity activity, and the volume of fat burned is greater in high-intensity activity (despite a lower proportion of fat meeting total energy requirements.) Therefore, athletes should work as intensely as possible within a given time frame to increase fat loss and optimize body composition.

Carbohydrate is a critical fuel for athletes because we can more efficiently create energy per unit of oxygen from carbohydrate than from any other fuel. One liter of oxygen can yield approximately 5 calories from carbohydrate but only 4.7 calories from fat. In addition, aerobic glycolysis can produce ATP for muscular work in a larger quantity and at a faster rate than the oxidation of fat can produce it. The increased energy efficiency of carbohydrate helps explain the muscular fatigue that quickly occurs during high-intensity activities when muscle glycogen is nearly depleted. We simply cannot supply sufficient ATP to working muscles to maintain the workload.

CENTRAL FATIGUE THEORIES

The conversion rate of adenosine diphosphate (ADP) to ATP is a critical step in the supply of energy to working muscles. Inadequate carbohydrate availability lowers the rate of ADP to ATP conversion, making it impossible for muscles to continue exercise at a high-intensity level. In addition, the failure to convert ADP to ATP causes a buildup of ADP that also contributes to muscle fatigue.

A number of other factors involving the central nervous system are also involved in muscle fatigue. As a group, these factors are referred to as *Central Fatigue Theory*. All of these theories involve mechanisms that cause more than the usual amount of the amino acid *tryptophan* to pass the blood-brain barrier, which stimulates an increase in the amount of serotonin (5-HT) that is produced. 5-HT is a neurotransmitter that causes people to feel relaxed and, if enough is produced, to feel sleepy. For the athlete, this could translate into muscular fatigue. Put simply, mental fatigue leads to muscle fatigue.

Theory 1: Low blood sugar and low muscle glycogen stores stimulate the muscle break down for gluconeogenesis. This results in an increased catabolism of branched-chain amino acids (BCAAs), which causes a reduction in circulating blood BCAAs. BCAAs and tryptophan compete for the same receptor carriers that enable their passage through the blood-brain barrier. When BCAAs are high, the tryptophan passage to the brain is controlled. However, when the BCAA blood level is low (as happens when they are catabolized for energy), tryptophan can sequester more of the receptor carrier, and more tryptophan enters the brain. Tryptophan stimulates the formation of 5-HT. To prevent this from happening, blood and muscle glucose levels must be maintained to avoid gluconeogenesis.

Theory2: Consumption of foods high in tryptophan (such as turkey) can increase the volume of tryptophan passing the blood-brain barrier, causing an increase in 5HT production. The increase in 5-HT leads to premature fatigue.

Theory 3: Fats compete for the same protein carrier in the blood as tryptophan. High fat intakes preferentially compete for this protein carrier, leaving a higher proportion of free tryptophan that can cross the blood-brain barrier. This causes an increase in 5-HT production, which may lead to premature fatigue. While it may be logical that the intake of both BCAAs and carbohydrate reduces 5-HT and, therefore, inhibits both mental and physical fatigue, studies have not been conclusive because of the difficulty in distinguishing between brain and muscle effects. In addition, there may be interference from compounds such as caffeine, the ingestion of which has been shown to delay fatigue by temporarily stimulating the central nervous system.

CARBOHYDRATE REQUIREMENTS

The Institute of Medicine recommends 130 grams (520 kilocalories) of carbohydrate per day, which is the average minimal usage of glucose by the brain. The desirable range of carbohydrate intake is 45 to 65 percent of total caloric intake (also referred to as the Acceptable Macronutrient Distribution Range, or AMDR), and the Daily Value (DV) for carbohydrate on food labels is based on a recommended intake of 60 percent of total caloric consumption. These recommendations also generally advise that no more than 25 percent of carbohydrate intake be derived from sugars (mono- and disaccharides).

Dietary fibre consumption (from indigestible and partially digestible polysaccharides) should be at the level of 38 grams per day for adult men and 25 grams per day for adult women. Adequate fibre consumption aids in the maintenance of normal blood sugar, reduces heart disease risk, and lowers constipation risk. The difference between genders in recommended fibre consumption is based on the lower total food mass and calories typically consumed by women.

It has been suggested that our only true requirement for carbohydrate is for vitamin C, a six-carbon, glucose-like substance that most animals can derive through an enzyme conversion of glucose. There is some historical evidence that our human ancestors consumed very little carbohydrate and survived. However, when considering athletes and the mountain of research demonstrating that carbohydrate is the limiting substrate in athletic performance, it becomes clear that human survival and human performance are entirely different issues. Athletes need carbohydrate.

Athlete requirements for carbohydrate are based on several factors. Athletes must consume enough carbohydrate:

- To provide energy to satisfy the majority of caloric needs;
- Optimize glycogen stores;
- Allow for muscle recovery after physical activity;
- Provide a well-tolerated source of energy during practice and competition; and
- Provide a quick and easy source of energy between meals to maintain blood sugar.

The traditional guideline for determining caloric intake has been to consider the amount of carbohydrate to be consumed as a proportion of total caloric intake. The recommendation for the general population is that carbohydrate should supply 50 to 55 percent of total calories, and the Dietary Reference Intake (DRI) is 130 grams per day (520 calories per day) for male and female adults.

However, the amount typically recommended for athletes is between 55 and 65 percent of total calories, assuming an adequate total caloric intake. Another, and clearly better, way of determining carbohydrate requirement is by taking into consideration the amount of carbohydrate to be consumed (in grams) per kilogram of body mass. The carbohydrate intake recommendations for endurance-trained athletes range from between 7 and 8 grams of carbohydrate per kilogram of body weight per day. Studies of carbohydrate consumption of different athlete groups have found differences in carbohydrate intakes.

These data suggest that an athlete should consume between 5 and 10 grams of carbohydrate per kilogram of body mass, or between 20 and 40 kilocalories of carbohydrate, per kilogram of body mass. A hypothetical 155-pound (70 kilogram) athlete would take in between 1,400 and 2,800 calories from carbohydrate, which represents a far greater carbohydrate consumption than the DRI of 520 calories. Assuming this represents approximately 60 percent of total calories from carbohydrate, this athlete would consume a total of 2,300 to 4,700 kilocalories per day.

Following the same logic, a 300-pound (136 kilogram) lineman on a football team would require 2,700 to 5,400 kilocalories just from carbohydrate each day, an amount that would be difficult to consume because carbohydrate has a relatively low energy density (i.e., only 4 kilocalories per gram). The generally recommended carbohydrate intake is based on the intensity and duration of exercise, with a higher requirement for greater duration and greater intensity. Most carbohydrates are derived from cereals, legumes, fruits, and vegetables. There is no discernable amount of carbohydrate in meats, and only a small amount of carbohydrate in milk and cheese. A number of dairy products have sugars added (yogurt, ice cream) to make them more widely acceptable.

The glycemic index is a measure of how quickly consumed carbohydrates manifest themselves as blood glucose. Foods are compared with the ingestion of glucose, which enters the blood quickly because it requires no digestion and is readily absorbed. Glucose has a glycemic index score of 100, which is the basis of comparison for other foods. Foods are compared on an isocaloric basis (i.e., all providing the same number of calories), which is logical but is also the cause of some of the confusion associated with the glycemic index. For instance, carrots have a high glycemic index (>85) but the amount of carrots typically consumed is so low that the total calories of glucose from carrots entering the blood would be small.

PROTEIN AND PHYSICAL ACTIVITY

Protein utilization is, to a large degree, a function of total energy intake adequacy. An inadequate total energy intake forces athletes to burn protein for energy, making less protein available for other critical functions. Therefore, the protein requirement for athletes (i.e., 12 to 15 percent of total calories or 1.2 to 1.7 grams per kilogram) is based on the assumption that total energy intake is adequate. A standard tenet in nutrition is that carbohydrate has a protein-sparing effect. This means that if you can supply sufficient carbohydrate to the system for fuel, then protein will be spared from being burned so it can be used for more important functions.

Studies have generally found that the maximal rate of protein utilization for nonenergy uses is approximately 1.5 grams of protein per kilogram of body weight. If this amount is exceeded, body tissues must make some decisions about what to do with the excess. The excess can be stored as fat, or some of the excess can be burned as energy. In either case, nitrogen must be removed from the amino acids, and this nitrogenous waste must be removed from the body. Virtually all studies that have looked at the total energy consumption of athletes indicate that athletes consume less total energy than they should to support the combined needs of activity, growth, and tissue maintenance. Since burning protein causes a lot of metabolic waste, it would be better to meet the energy requirement by providing a cleaner-burning fuel-carbohydrate. A goal for most athletes is to remain in a nitrogen-balanced state in which as much nitrogen is coming into the system as is being excreted. A negative nitrogen balance suggests that more nitrogen is being excreted than is being consumed, a state that will inevitably lead to muscle loss.

A positive nitrogen balance suggests that more nitrogen is being retained than is being excreted, a state that suggests that muscle is being gained. The amount of protein required to maintain a nitrogen-balanced state in nonathlete adults has been well studied and established at the level of .8 grams of protein per kilogram of body weight per day, while for athletes an intake of between 1.2 and 1.7 grams of protein per kilogram of body weight per day is needed. Both the athlete and nonathlete recommendations are based on the consumption of a total caloric intake that satisfies energy needs.

The higher protein recommendation for athletes is based on four factors: (1) Athletes typically have a higher lean body mass that requires more protein to sustain; (2) athletes lose a small amount of protein in the urine, and nonathletes do not (the greater the intensity and duration, the greater the proteinuria); (3) athletes "burn" a small amount of protein (approximately 5 percent of total energy combustion) during physical activity; and (4) athletes require additional protein to recover from the muscle damage that occurs during training .

The Institute of Medicine has stated that additional protein for healthy adults who exercise regularly is not needed because exercise increases protein retention. Nevertheless, both the American College of Sports Medicine and the American Dietetic Association recommend that protein intakes range between 1.2 and 1.7 grams per kilogram of body weight in physically active people. In reality, most athletes consume far more protein than they require, and typically more than the maximum recommended level of 1.7 grams per kilogram. Some strength and power athletes regularly consume 300 to 775 percent of the recommended level for protein (.8 grams per kilogram). Possible exceptions to high protein intakes may occur among vegetarian athletes and athletes in subjectively judged sports who strive to maintain low body weights (i.e., gymnasts, divers, figure skaters).

Protein oxidation during short-term, intense exercise is insignificant, but protein provides from 3 to 5 percent of total energy needs during endurance exercise. Protein utilization during exercise will rise to a level greater than 5 percent of total energy needs if glycogen levels are low, blood sugar is low, exercise intensity is high, or exercise duration is long. There is a common misunderstanding that extra protein intake alone will support a larger muscle mass, and this theory is the main rationale for the large protein intakes seen in many athletes. In fact, additional total calories are required to support a larger muscle mass, and protein should constitute the same relative proportion of the extra calories consumed.

For instance, if a 75-kilogram (165 pound) man wishes to increase his muscle mass by 3 kilograms (6.6 pounds), he would need to consume approximately 1.5 additional grams of protein for each kilogram of muscle mass desired. This amounts to only 4.5 grams of additional protein to support the larger muscle mass. By contrast, 30 grams per kilogram of additional carbohydrate, or 90 grams of additional carbohydrate in total, is required to support the larger muscle mass. Here is the total additional caloric requirement represented by the additional muscle:

- 4.5 grams protein × 4 calories per gram = 18 kilocalories from protein
- 90 grams carbohydrate × 4 calories per gram = 360 kilocalories from carbohydrate
- Total additional calories = 378 calories per day above current requirements to support a 3-kilogram increase in muscle mass

Of course, this athlete would also need to stimulate muscle enlargement by undertaking the appropriate strength-building exercises. Otherwise, the extra calories would manifest themselves as stored fat rather than additional muscle. It is likely that the large amount of protein consumed by so many athletes represents the extra calories they require to maintain or enlarge the muscle mass. Although it is certainly possible to use protein as a primary energy source, it is not the most desirable source because of the nitrogenous wastes produced with protein oxidation. In addition, protein can be an expensive source of calories when provided in supplement form. For instance, eggs (an extremely high-quality source of protein) cost approximately 13 cents per 8 grams of protein, while protein capsules cost approximately $1.20 per 8 grams of protein and may be of questionable quality.

High-protein foods have a long gastric emptying time so are not recommended immediately before or during exercise. In addition, there is no evidence that adding protein to a glucose- and sodium-containing sports beverage does anything useful for either endurance or power enhancement. In fact, protein added to a sports beverage that is consumed during competition increases the risk of gastrointestinal distress and may delay the delivery of fluids and carbohydrate to needy muscles. Protein added to a sports beverage reduces the content of what athletes really need: fluid, carbohydrate, and electrolytes. Therefore, the majority of energy in the preexercise meal and during exercise fluid replacement should be from carbohydrate. An increasing body of evidence suggests that adding small amounts of protein to postexercise food and drink is useful for muscle recovery, although the benefit of protein is reduced if sufficient carbohydrate is ingested postexercise to replenish glycogen stores. Consuming a drink that contains approximately .1 gram of protein per kilogram (30 calories of protein for a 75-kilogram athlete) after heavy resistance training does appear to improve muscle protein balance. The general guideline is for endurance athletes to consume carbohydrate at the rate of a minimum of 1.2 grams per kilogram of body weight each hour during the first 3 to 5 hours after prolonged exercise. This strategy ensures muscle glycogen replenishment. The addition of some protein in this carbohydrate mix may be useful for muscle recovery, but the majority of postexercise substrate should clearly be from carbohydrate.

FLUIDS AND ELECTROLYTES

Perhaps the single most important factor associated with sustaining a high level of athletic performance is maintenance of fluid balance during exercise. Despite this, most athletes experience deterioration in hydration state (with a resultant drop in blood volume) during training and competition. Studies demonstrate that, even in the presence of available fluids, athletes experience a degree of voluntary dehydration that has an inevitably negative impact on performance. Given the tremendous amount of heat that must be dissipated during exercise through sweat evaporation, athletes have no reasonable alternative for sustaining exercise performance other than to pursue strategies that can sustain the hydration state.

A failure to do so will result in premature fatigue and may also lead to potentially life-threatening heatstroke. This chapter discusses the strategies related to achieving and sustaining an optimal hydration state and reviews studies that have assessed the optimal concentration of carbohydrate and electrolytes for the "ideal" sports beverage. Water is the main component of blood, which delivers oxygen, nutrients, hormones, and a multitude of other substances to cells and removes metabolic by-products from cells. Water also has a protective function, cushioning the spinal cord and brain from sudden-impact injury, and is a critical component of our temperature regulation mechanism. Water and its electrolyte components are involved in the control of osmotic pressure, regulating the amount of fluid inside and outside cells. Well-hydrated athletes are referred to as being euhydrated or normohydrated; those with below-normal body water levels are referred to as hypohydrated or, if severe, dehydrated; and those with above-normal body water levels are referred to as hyperhydrated. We have systems for the normal control of body water levels, involving an increased retention of body water or an increased loss of body water, all mediated through a series of hormones stimulated by osmoreceptors that monitor blood osmolality and volume receptors that monitor the volume of extracellular water. Excretion of fluids and metabolic by-products is a main function of the kidneys, which are stimulated by hormones and enzymes to adjust the volume of water and electrolytes excreted or retained. The concentration of sodium is a primary influence on the osmolality of extracellular fluid, which is maintained within a narrow range. Because sweat is hypotonic, prolonged exercise results in a higher plasma osmolality (more water is lost than sodium). As a means of preserving body water volume, urine production during and shortly after exercise is slightly decreased. If the blood has a relatively high concentration of sodium, protein, or glucose per unit volume of fluid (i.e., is hypertonic), water is drawn from cells to normalize the concentration of electrolytes. Receptors in the hypothalamus detect the fact that the blood is hypertonic through its osmoreceptors, which leads to the release of antidiuretic hormone (ADH) from the pituitary gland. ADH forces the kidneys to reabsorb more water by producing a more concentrated urine.

For this reason, a common test for adequate hydration status is urine colour, with dark urine indicating a greater degree of underhydration than light urine. The osmoreceptors can also induce the sensation of thirst, although this sensation rarely occurs before the loss of 1.5 to 2.0 liters of water (1 liter equals approximately 1 quart;. Since it is nearly impossible for athletes to consume sufficient fluids during physical activity to maintain the body water level, waiting for the thirst sensation before drinking fluids guarantees that the athlete will be exercising in a progressively worsening state of underhydration.

Where's the Water?

- 66% of a person's total body weight is from water.
- 65% of total body water is intracellular.
- 35% of total body water is extracellular.
- Well-hydrated muscles are about 75% water.
- Bones are about 32% water.
- Fat is essentially anhydrous, having only about 10% water content.
- Blood is about 93% water.
- Average males are about 60% water weight.
- Average females are about 50% water weight.
- Obese individuals are about 40% water weight.
- Athletes are about 70% water weight.

Note: The higher the musculature and the lower the body fat, the higher the contribution of body water to total body mass.

Common Conversions

- To convert Fahrenheit to Celsius, subtract 32 degrees and divide by 1.8.

- To convert Celsius to Fahrenheit, multiply by 1.8 and add 32 degrees.
- To convert quarts to liters, multiply quarts by .946.
- To convert liters to quarts, multiply liters by 1.057.

In a state of hyperhydration, the concentrations of electrolytes, protein, and glucose are lower than normal in the blood. This condition shuts down the production of ADH so that diluted urine is produced. Fluid tends to migrate from blood to cells to adjust for this hypotonic state.

Blood volume is affected by the concentration of sodium, the main extracellular electrolyte. A high sodium concentration is associated with an eventual enlargement of the blood volume, which results from the body's attempt to normalize the concentration of sodium per unit of fluid volume. The reverse situation, a low sodium concentration, is typically associated with an eventual reduction of the blood volume. To adjust for the natural variations in sodium intake, the hormone aldosterone is produced to retain more sodium in a low-sodium environment, and aldosterone production ceases when sodium concentrations are high so as to cause the excretion of more sodium.

Under normal circumstances, the combination of volumetric controls, osmo receptors, antidiuretic hormone, and aldosterone maintains a relatively steady blood volume even with variations in fluid and sodium consumption. Exercise leads to an increased production of ADH and aldosterone, both of which conserve body water and sodium. This system is sufficiently effective that fluid deficiencies leading to physiological problems are rare, even in athletes. However, exercise at high intensity or of long duration (or both), particularly in a hot and humid environment, places the athlete at hydration risk because fluid loss (through sweat) may exceed the athlete's capacity to consume and absorb fluids. This can lead to a progressive reduction in blood volume, a reduced sweat rate, and other problems that negatively affect performance and health; therefore, it is important for athletes to always maintain fluid balance.

Benefits of Maintaining Fluid Balance

Maintaining fluid balance during exercise helps sustain athletic performance through the following:

- Attenuation of increased heart rate
- Attenuation of increased core temperature
- Improvement in stroke volume
- Improvement in cardiac output
- Improvement in skin blood flow
- Attenuation of higher plasma sodium, osmolality, and adrenaline
- Reduction in net muscle glycogen usage

A Balance of Fluid Loss and Intake

Physical activity creates heat, and this heat must be dissipated for the athlete to continue performing the activity. Failure to dissipate heat will eventually lead to heatstroke and, potentially, death. One of the main mechanisms for dissipating heat is sweat production; sweat cools the body down when it evaporates off the skin. The inability to produce sufficient sweat will cause the body to overheat. Since athletes have a finite storage capacity for water, and a tremendous ability to produce sweat, fluids must be consumed during physical activity to maintain the sweat rate.

Athletes working intensely in the heat can lose 2.5 liters of sweat per hour. Sweat contains electrolytes (mainly sodium chloride but also potassium, calcium, and magnesium), with a sodium concentration that ranges from 20 to 80 millimoles per liter, depending on common sodium consumption in the diet, the sweat rate, acclimatization to the heat (better acclimatization results in lower sodium loss), and content and amount of rehydration beverages.

Because the electrolyte concentration of sweat is different from that of plasma and intracellular water, there are concerns that an electrolyte imbalance will develop with intense physical activity. Of greatest concern is the potential sodium imbalance that could occur. The loss of a single liter of sweat containing 50 millimoles per liter of sodium translates into a loss of nearly 3 grams of sodium chloride. Athletes who lose 2.5 liters of sweat per hour will lose almost 15 grams of sodium in 2 hours, a level that could easily exceed normal daily sodium intakes.

Temperature regulation represents the balance between heat produced or received (heat-in) and heat removed (heat-out). When the body's temperature regulation system is working correctly, heat-in and heat-out are in perfect balance, and body temperature is maintained. Both internal and external factors can contribute to body heat. Radiant heat from the sun contributes to body temperature, as does the heat created from burning fuel (carbohydrate, protein, or fat). Somehow, athletes must find a way to dissipate from the body the same amount of heat that has been added to the body to maintain a constant body temperature.

The two primary systems for dissipating, or losing, heat involve:

1. Moving more blood to the skin to allow heat dissipation through radiation
2. Increasing the rate of sweat production.

These two systems account for about 85 percent of heat removal when a person is at rest. Heat losses through conduction (the natural transmission of heat from a hotter body to the cooler air environment) and convection (heat transfer from tissue to the blood and through the skin) account for the remaining 15 percent of heat-out. During exercise, however, virtually all heat loss occurs via evaporation (sweat). Both of these systems rely on maintenance of an adequate blood volume. A lower blood volume results in a reduced movement of blood to the skin, and sweat production is also reduced.

Working muscles demand more blood flow to deliver nutrients and to remove the by-products of burned fuel. However, at the same time there is a need to shift blood away from the muscles and toward the skin to increase the sweat rate. With low blood volume, one or both of these systems fail, with a resultant decrease in athletic performance. In fact, the maintenance of blood volume is rightly considered by many to be the primary indicator of whether an athlete's performance can be maintained at a high rate.

Energy metabolism is only about 20 to 40 percent efficient, meaning that only 20 to 40 percent of food energy can be converted to the mechanical energy of muscular work. The remaining 60 to 80 percent of the food energy that is burned is lost as heat. However, when the rate of energy burn goes up, as happens during physical activity, the amount of heat added to the system is dramatically increased, so the heat-out systems must be "turned up."In fact, heavy exercise can produce 20 times the amount of heat produced at rest. Without an efficient means of heat removal, body temperature will rise quickly. The upper limit for human survival is about 110 degrees Fahrenheit (43.3 degrees Celsius), or only 11.4 degrees Fahrenheit (or 6.3 degrees Celsius) higher than normal body temperature. Body temperature has the potential to rise approximately 1 degree Fahrenheit every 5 minutes. It is conceivable, therefore, that an underhydrated athlete could be at risk for heatstroke and death less than 1 hour after the initiation of exercise.

Athletes doing very mild exercise that burns 300 kilocalories of energy during 30 minutes would use approximately 75 kilocalories for muscular work, and 225 kilocalories would be lost as heat. This excess heat must be dissipated to maintain normal body temperature. Athletes working twice as intensely would create 450 kilocalories of excess heat that would need to be dissipated over the same 30 minutes to maintain body temperature. It is estimated that 1 milliliter of sweat can dissipate .5 kilocalories, so over this 30-minute period the athlete would lose approximately 900 milliliters (almost 1 liter) of sweat. In 1 hour of high-intensity activity, approximately 1.8 liters of water would be lost. On sunny and hot days when the heat of the sun is added to the heat generated from muscular work, the athlete must produce more sweat to remove more heat. The fluid requirement is compounded when exercising intensely on a hot and humid day Sweat doesn't evaporate off the skin as easily when it is humid, so even more sweat must be produced. In these conditions, a person can easily lose between 1 and 2 liters of fluid (via sweat) per hour.

Well-trained athletes exercising in a hot and humid environment may lose more than 3 liters of fluid per hour. To protect athletes from placing themselves at increased heat-stress risk, the heat index was developed. This index simultaneously considers environmental temperature and relative humidity to establish exercise risk.

Factors Affecting Fluid Loss

Because sweat has a lower osmolality than does plasma (i.e., sweat is hypotonic), profuse sweating increases plasma osmolality. Whether or not this increased plasma osmolality affects body temperature or cooling capacity in an exercising individual is, as yet, unclear, but a sufficient change in osmolality and volume does stimulate the kidneys to excrete sodium and reduce urine output by producing more concentrated urine. Several factors affect the rate at which an athlete can produce sweat. Higher ambient temperatures result in a greater potential for sweat production. Higher humidity is also responsible for higher sweat production, but because the vapor pressure gradient and skin is low, the cooling potential (i.e., the rate of evaporation off the skin) is lower in humid environments. The same problem also exists with clothing that traps sweat against the skin (i.e., does not breathe). This type of clothing results in a reduced cooling efficiency that forces a greater sweat rate. (Sweat-soaked clothing doesn't mean an athlete is effectively controlling body temperature, it just means he or she is losing water.)

Some new materials designed for athletes actually wick sweat away from the skin to improve evaporative efficiency. Athletes with large body surface areas may also have an enhanced sweat production capacity and, therefore, an enhanced evaporative heat loss. But these athletes may also gain more heat from the environment through radiation and convection in hot weather. The conditioning or training state of an athlete makes a difference. Well-conditioned athletes have a higher sweat volume potential that results in an enhanced cooling potential. However, this higher sweat rate requires a greater during-exercise fluid consumption to avoid higher heat-stress risk. An athlete's state of fluid balance also plays a factor. The better the hydration state, the greater the sweat potential.

As athletes become progressively dehydrated, the sweat rate is reduced, and body temperature rises. This is a problem because fluid consumption during activity is rarely greater than 2 cups (480 milliliters) per hour, or only 30 to 40 percent of the amount of fluid lost in sweat, an amount that will inevitably lead to the athlete's becoming dehydrated. Consider that marathoners competing in a cool temperature of 50 to 54 degrees Fahrenheit (10 to 12 degrees Celsius) lose between 1 and 5 percent of total body mass. Marathoners competing in warm weather lose about 8 percent of total body mass, or between 12 and 15 percent of total body water.

Factors Affecting Fluid Intake

The two main factors influencing fluid intake are thirst and taste. Thirst is a sensation of dryness in the mouth and throat related to the body's need for additional fluids. Taste is the response humans have (either good or bad) to substances in the mouth. Humans are more likely to consume more of what they like, or more of what tastes good to them. Most athletes induce "voluntary dehydration" because they don't drink enough despite having plenty of fluids readily available. Insufficient consumption of fluids by athletes is probably due to a lack of the thirst sensation. The onset of thirst may be the result of habit, ritual, or the need for a warming (hot fluids) or cooling (cold fluids) effect. A rise in plasma osmolality of between 2 and 3 percent is needed to produce the sensation of thirst, and sensitivity to a reduction in fluid volume is even less responsive, requiring nearly a 10 percent decrease in blood volume to stimulate thirst.

The thirst sensation, often considered delayed in athletes because it doesn't appear until an athlete has already lost 1.5 to 2.0 liters of body water, is therefore a poor indicator of fluid needs in athletes. There is no hope that an athlete can return to an adequately hydrated state during exercise if fluid consumption begins at the same time the thirst sensation occurs. This apparent delay in the thirst mechanism is a primary reason for athletes to train themselves to consume fluids on a schedule, whether they feel thirsty or not.

Colour, taste, odor, temperature, and texture all play important roles in determining if the beverage will be considered desirable and whether it will be consumed. It appears that athletes prefer cool beverages with a slightly sweet flavour. Heavily sweetened beverages (around a 12 percent carbohydrate solution) are not as widely tolerated during exercise as beverages with a 6 or 7 percent carbohydrate solution.

When not exercising, however, the reverse may be true, pointing to an interesting phenomenon of exercise: Food and drink taste differently while exercising than when not exercising. Therefore, athletes are wise to determine the fluids they find most desirable for exercise while they are exercising.

CONSIDERATIONS FOR ENERGY INTAKE

The question most frequently asked by athletes concerns what to eat before competition. Although this is important, it is of relatively small importance when compared with how the athlete should eat most of the time. It is impossible to properly prepare for a competition by consuming some pancakes several hours before the feet are placed in the starting blocks. It takes a consistent and long-term effort in conditioning and good nutrition.

There is no way an athlete with iron deficiency can magically cure the condition by consuming some red meat the day before an event. It may take 6 months on a proper diet to reach a state of normal iron status. Therefore, the first and most important step in preparing for competition is to consistently eat enough energy and nutrients to support the body's energy and nutrient requirements. Failure to do so will inevitably lead to a poor competition outcome, no matter what you do just before the competition.

In addition to consuming enough energy and nutrients, it's equally important to eat foods when the body can benefit the most from them. The timing of meals is also important to make certain the muscles have enough energy and nutrients to grow and get stronger during training sessions rather than get burned for energy because the athlete hasn't eaten enough. Put simply, it's important to get enough and get it on time. This isn't easy to accomplish because athletes have terribly hectic schedules, and it takes strategic thinking and good scheduling to ensure food is consumed when it's needed. Although careful meal planning may not seem as important as having a well-developed training plan, both should be considered equally important. They should also be thought of collectively to make certain the training plan can be properly supported with the foods that are consumed.

If the general food intake is supportive of the training plan, what should an athlete do differently on the days leading up to a competition? The sequence of events for the seven days before a competition should meet three major goals:

- The athlete should gradually become rested. This may be a problem for many athletes and coaches because athletes (either with or without the encouragement of the coach) often increase the training schedule during the week leading up to a competition. Overtraining is a big problem and may increase the risks of getting sick or getting injured. It certainly doesn't help an athlete do his or her best at the upcoming competition.
- The athlete should gradually build up muscle glycogen (energy) stores. The main purpose of gradually reducing the intensity and duration of training sessions before the competition is to be sure the athlete can begin the competition with full muscle glycogen stores. The storage capacity for glycogen is relatively small, and athletes are heavily reliant on stored glycogen for muscular work (it's the limiting fuel for muscular work, regardless of the type of exercise the athlete is doing). Therefore, it's important to eat plenty of carbohydrates and reduce work so glycogen stores are full going into the competition.
- The athlete should become well hydrated. When athletes work hard it is difficult (if not impossible) to maintain an optimal hydration state. It takes time to return lost body water, and athletes should give themselves the pportunity to do so by reducing the training intensity and duration and by drinking plenty of fluids. An additional benefit of becoming well hydrated is that glycogen storage is enhanced. The gradual tapering of training during the 7 days before competition makes it easier for the athlete to start the competition in a well-hydrated and optimally energized state.

Of course, many sports don't provide athletes the luxury of tapering activity on a 7-day cycle. Basketball and hockey players play several games each week during the season, and baseball players play nearly every day. Although their schedules don't permit 7-day activity tapering, the principles behind tapered activity, glycogen storage, and optimal hydration should be remembered and, when possible, adhered to. For athletes with daily schedules that eliminate the possibility of tapering, consumption of high-carbohydrate diets and maintenance of optimal hydration become even more important components of athletic performance. Athletes with these schedules should develop eating and drinking plans that are as solid as their training and competition plans. All too often athletes prepare for a big competition by increasing their training regimen as the competition draws nearer. This is a big mistake. Coaches working in high-skill sports, such as figure skating and gymnastics, may ask their athletes to perform multiple run-throughs of their routines the day before competition just to be sure they can do them. The message this sends to an athlete (i.e., "I don't believe you've ready, and we've going to keep practicing until you get it right") is counterproductive.

There is nothing more confidence building for athletes than entering the competition well rested and knowing the coach is secure in their ability to do a good job. This is true whether an athlete is a professional or a tee ball player in little league baseball.

Carbohydrate Ingestion Before Exercise

A high-carbohydrate meal that is completed approximately 90 minutes before physical activity has been shown to improve endurance performance. After this preexercise meal, athletes should consume carbohydrates right up to the beginning of the training session or competition to avoid low blood glucose.

Two strategies can be followed:

- Ingestion of a carbohydrate-containing sports beverage using a sipping strategy, where approximately 2 to 4 ounces (60 to 120 milliliters) of beverage is consumed every 10 to 15 minutes.
- Snacking on low-fibre, starchy foods (such as saltine crackers) every 15 minutes, washed down with ample quantities of water.

Athletes should avoid eating patterns that could stimulate a reactive hypoglycemia, which is caused by ingestion of large quantities of foods with a high glycemic index, but should also avoid the hypoglycemia that could occur from consuming no carbohydrates or from delayed eating. Snacking and sipping procedures appear to be well tolerated and help maintain blood glucose.

Carbohydrate Maintenance During Exercise

Avoiding low blood glucose and avoiding depletion of muscle glycogen stores are both critical for maintaining exercise performance. Consumption of carbohydrate-containing beverages (e.g., sports beverages) and food during exercise delays fatigue and improves performance, even if the consumption occurs late in the exercise session.

This strategy delays fatigue through the following mechanisms:

- Maintains blood glucose, which preserves liver glycogen
- Maintains branched-chain amino acid (BCAA) levels, which avoids central fatigue by maintaining the ratio of tryptop an and BCAAs

- Inhibits the production of cortisol, which is catabolic to muscle tissue
- Reduces the usage of muscle glycogen by providing a constant source of glucose from the blood to working muscle cells

During exercise, carbohydrate is best obtained through a 6 to 7 percent carbohydrate solution, with 4 to 8 ounces (120 to 240 milliliters) taken every 10 to 20 minutes (the amount to consume depends on sweat rate). A number of different carbohydrate solutions are available to athletes, each with a different concentration and composition of carbohydrate. Issues of gastrointestinal distress and osmolarity should be considered. Of equal 6 percent concentrations of glucose, fructose, or sucrose, the fructose has been shown to cause more gastrointestinal distress. Therefore, athletes should carefully check their tolerance of fructose-only beverages before consuming them in critical situations. (Most sports beverages contain multiple carbohydrate types.)

Few sports beverages are milk based or use lactose as a predominant form of carbohydrate because of the relatively common problem of lactose intolerance. This condition, caused by an inadequate production of the enzyme lactase, results in diarrhea, gas, and abdominal pain. Given the possibility of lactose intolerance in some athletes, it is prudent for athletes to avoid lactose-containing products immediately before and during physical activity.

Glucose polymers have the advantage of being able to deliver more carbohydrate in a lower osmolar solution. Isocaloric with simple sugars, glucose polymers improve gastric emptying and enhance absorption. Athletes involved in extremely high-intensity activity for a long duration may need a high volume of carbohydrate calories during physical activity, and glucose polymers may provide a good solution for these athletes.

Carbohydrate Replenishment After Exercise

Glycogen and fluids are usually, to a degree, depleted after exercise, and protein requirements are also higher to aid in muscle recovery. The protein and fluid issues are discussed in other sections of this book. The carbohydrate and glycogen issues are presented here. One of the main postexercise goals is to replenish glycogen to prepare the athlete for the next bout of exercise. As glycogen becomes depleted, the enzyme glycogen synthetase becomes elevated in the blood.

Providing glucose or sucrose (but not fructose) while glycogen synthetase is elevated efficiently replaces muscle glycogen stores. Glycogen synthetase reaches its peak at the point of greatest glycogen depletion, which is immediately after exercise. Therefore, athletes should consume carbohydrates as soon as physical activity ends. Ideally, the carbohydrate consumed for the first 2 hours after exercise should be high glycemic, followed by medium-glycemic carbohydrates for another 2 hours and finally medium to high-glycemic carbohydrates for the remainder of the day. Athletes should plan on consuming 200 to 400 kilocalories from carbohydrate (50 to 100 grams) immediately after physical activity, followed by sufficient carbohydrates to fulfill the guidelines.

Carbohydrate Loading

Consumption of carbohydrates before exercise improves carbohydrate stores and reduces the chance for premature fatigue, regardless of whether the sport is high endurance and low intensity, intermittent (as in many team sports), or high intensity and low endurance. Carbohydrate loading is a strategy commonly followed by many athletes before an endurance competition to increase muscle glycogen storage. The general technique is to gradually increas carbohydrate and fluid intake each day, beginning the week before competition, while exercise is tapered downward. This reasonable, safe strategy maximizes glycogen storage.

An older strategy for carbohydrate loading involved depleting carbohydrates by exercising intensely while consuming low-carbohydrate foods.

This was followed by the technique described in the previous paragraph. This older carbohydrate-loading technique is dangerous (depletion of glycogen stores may cause a sudden and dangerous drop in blood pressure), and there is no evidence that it better optimizes glycogen stores.

Seven-Day Taper

The following tables provide an example of how the principles for maximizing carbohydrate storage can be put to work. They illustrate what and how athletes might eat if they typically train twice daily. You'll notice that the food is spread out over six smaller meals rather than two or three larger ones. You'll also notice that the caloric level of the meals does not emphasize dinner at the end of the day. Although dinner is important, training takes place before dinner, so ample energy must be available when the athlete needs it the most.

Breakfast comes before the morning workout; when an athlete wakes up, blood sugar is marginal and the liver is virtually depleted of energy, so maintaining blood sugar is virtually impossible. Eating some food before the morning workout ensures that the muscles will benefit from the training and makes the athlete feel better. Nobody feels good with low blood sugar. As indicated by the tables, foods should be consumed long enough before the training session so an athlete feels no discomfort from training because there is food in the stomach. In addition, the meal plan always includes some carbohydrates immediately after a workout. This helps ensure an effective replenishment of the glycogen that was used up during the training. Waiting too long after training to eat can diminish the efficiency of muscle glycogen replacement.

7

Enzymes and Food Processing

INTRODUCTION

Food processing enzymes are used as food additives to modify food properties like digestability, texture and shelf life. Major enzymes used in the food industry are for starch liquefaction, saccharification and isomerization reaction. Food Processing enzymes are used in meat processing, dairy industry and in manufacture of pre-digested foods. Enzymes are also used for natural breaking of large sized molecules of carbohydrates, fats and proteins, in fermentation and preparing predigested protein food or preparing amino-acids and specialty foods. The project discussed in this chapter envisages setting up of a food processing enzyme unit in the state of Gujarat.

MARKET AND GROWTH DRIVERS

The global market for food processing enzymes in 2002 was approximately US $700 million that increased to US $740 million in the year 2004. Food and feed enzymes together had nearly 45 per cent share, followed by industrial enzymes which are used in the detergent industry.

Table. Sales INR in Millions

Sr. No.	Name of Company	2003-2004	2004-05	Growth (%)
1.	Biocon Ltd.	5020.0	6464.0	29
2.	Novozymes Bangalore	530.0	690.0	30

In terms of value, this is contributing approx. US $450 million in the world enzyme market estimated at US $1000 million by the year 2005. However, food processing enzymes find limited applications in India and hence its market also has remained un-tapped till date. Enzymes for food and feed processing industries contribute almost 17 per cent to the Indian market. In recent years, the trend is changing with globalization of food processing industry in India, Food processing enzyme market is fast growing in diverse applications. Biocon Ltd and Novozymes are the two prominent players in the Indian food processing enzymes industry, their sales turnover for 2003-04 and 2004-05 and growth rate is summarized here:

GROWTH DRIVERS

- Population growth, urbanization and rising income levels have increased demand on food items with its spill-over effect seen on the growth of demand of food products and this in turn has spurred the demand of food enzymes in India.
- Technical innovation in enzymatic food processing is a major growth driver for this segment.
- The fast changing food habits, increased use of fast foods, modified foods and instant foods are other drivers for demand of food processing enzymes.
- Health food is a sunrise industry in India and also globally, where food processing enzymes find wider application in manufacturing of these products.

TECHNOLOGY/PROCESS

Fermentation

- Fermentation of a selected microorganism under sterile conditions.
- Production of the desired enzyme by consumption of carbohydrates, proteins, salts, water and energy
- Transfer of the fermented broth to recovery.

Recovery

- Filtration of the fermentation broth.
- Purification of the liquid phase containing the enzyme.
- Concentration of the liquid enzyme to desired enzymatic activity.
- Stabilization and standardization of a liquid product.
- Standardization of a liquid concentrate for granulation.

Granulation

- Formulation and standardization of the enzyme into solid granules to encapsulate the enzyme.
- Providing the right property to the product.
- Preparation of source materials for enzyme extraction of food processing enzymes using physical/ chemical processes.
- Production of food processing enzyme using physical/ chemical/bio-tech processes.
- Purification process quality control checks packing in air tight bulk packing.

Food processing enzymes are produced, mostly using all natural materials as source of raw materials. Now, these enzymes are produced in bulk using fermentation techniques-both with solid state and submerge culture in modern process of manufacturing.

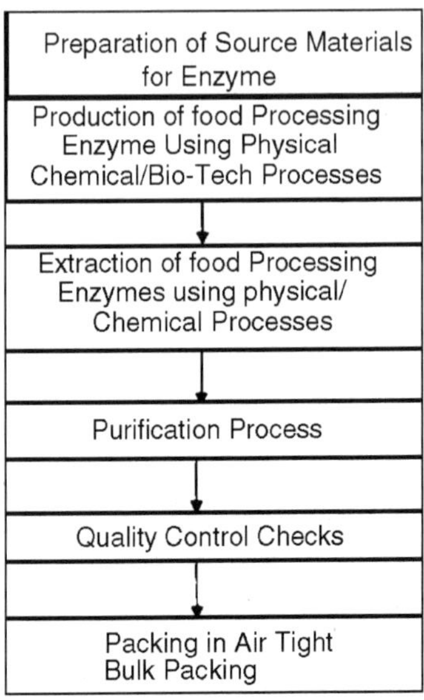

RAW MATERIALS

In the following Table 2 particulars of various food processing enzymes, their source and applications are summarized:

Table. Food Processing Enzyme—Source and Their Application

Sr. No.	Enzyme	Source	Applications
1	L-amylase, Bacterial	*B.subtillis/ B.licheniformis*	Starch conversion
2	L-amylase, fungal	*Aspergillus oryzae*	Maltogenic Saccharufucatuib
3	Amylogluco-sidedase	*Aspergillus nigar var,*	Starch syringes dextrose, foods
4	B-Glucanase	*B.subtillis/ Peni-cillium emersonil*	Brewing and food processing
5	Neutral protease	*B.subtillis*	Brewing/ flavouring
6	Neutral protene	*Aspergillus* oryzae	Baking
7	Cellulase	Trochoderma Spp	Cellulose hydrolysis
8	Invertase	Yeast SPP	Confectionary
9	Pectinase	*A.niger*	Fruit/wine processing
10	Papain	*Papaya latex*	Meat processing

11	Rennet	Mucor Spp	Dairy industry
12	Glucose isome-	Strep Spp	High fructose corn syrup
13	Lipasa	*Mucor arase* Spp/ Aspergillus	Dairy Indus/ Fat splitting
14	Lactase	*S.Lactis*	Dairy industry
15	Hemicellulase	*A.niger*	Baking, fruits, gums
16	Glucose oxidase	*A.niger*	Analytical, food processing
17	Catalase	*A.niger*	Analytical, food processing

SUGGESTED PLANT CAPACITY AND PROJECT COST

Capacity – 360 Tonnes per annum.
Capital cost is estimated to be INR 50 million.

Table. Estimated Project Cost and Means of Finance

Sr. No.	Cost of Project	INR in Millions
1	Land and Land Development	3.60
2	Building and Civil Works.	6.75
3	Plant and Machinery	23.00
4	Misc. Fixed Assets	3.10
5	Preliminary and Pre-operative	4.00
6	Provision for Contingencies	2.70
	Total Fixed Assets	43.55
7	Margin Money for Working Capital	6.45
	Total	50.00
	Means of Finance	
8	Promoters Contribution	16.50
9	Term Loan	33.50
	Total	50.00

The proposed project will require approx 6,000 sq. mt. of land with an proposed built up area of 1,500 sq. mt. The unit is proposed to have an installed capacity of 360 TPA. The total fixed cost of the project is estimated at INR 43.55 and INR 6.45 million is the working capital margin which adds up to block capital cost of INR 50.00 million. The unit being proposed to cater to domestic as well as international demand is suggested to have a debt-equity ratio of 2:1. Thus, the estimated term loan amounts to INR 33.50 million and equity at INR 16.50 million.

PLANT AND MACHINERY

The list of main plant and machinery and utility equipments is as mentioned below:

UTILITIES

The unit will require utilities in the form of electricity and low pressure steam from non-IBR steam generating mini boiler. The unit will require approx. 220 KVA power and approx. 500 kg steam per hour for the proposed installed capacity of 360 MT/ per annum.

Table. List of Main Plant and Machinery

Sr. No	Particulars	Quantity	Supplier
1.	Preparatory for Enzyme source material including Sifter and particle classifier Enzyme Production	1 Set	Goldin India equipment Pvt. Ltd., Vadodara Pulverizer, Riddhi Pharma Machinery Ltd., Mumbai
2.	Fermentor for	2	Alfa-Laval India Enzyme 5000 Letre Ltd., Pune
3.	Glasslined reactors for Enzyme separation	2	Gammon India Ltd., Mumbai
4.	Steam Distillation Works, Mumbai	2	Dipesh Engineering Column Enzyme Extractor
6	Solvent extraction	1	Desmet Chemfoods equipments India Pvt.Ltd., Mumbai
7.	Enzyme separation	1	Ion Exchange India column Ltd., Mumbai
8.	Basket centrifuges	3	Sukhras Machines for final product Pvt. Ltd., Mumbai separation
9.	Hot air dryer with hot air generator for final product drying	1	Aerotherm India P.Ltd., Ahmedabad
10.	Bag filling and sealing	3	SPA International, Machines Deonar, Mumbai
11.	Oil fired steam generation unit	1	Thermax Ltd., Pune
12.	HT/LT Electrification standby		Lot Kirloskar Electricals and DG set for Ltd, Mumbai power

MANPOWER

The proposed project will have manpower requirement of 35 persons which is of two types, technical and non-technical. The proposed food processing enzyme unit will require 4 functional managerial persons, 1 personnel cum administrative manager, 10 technical and 15 non-technical persons, 5 office staff.

SUGGESTED LOCATION

The preferred location for the proposed project is Central and South

Gujarat areas close to urban centres, like Ahmedabad, Gandhinagar, Vadodara, Surat and Valsad districts.

PROJECT TIME LINE

The proposed project will have project time line of 5 to 6 months for obtaining various clearances and it will require 10- 12 months for implementation subject to availability of required permission, approvals and resources.

FINANCIAL INDICATORS

Based on the profitability projections worked out for the proposed project, key financial indicators are as summarized below:

Table. Key Financial Indicators

Sr. No	Financial Ratios	1st Year	2nd Year	3rd Year
A	Break-Even Point in% capacity	42.97	38.84	35.47
B	Debt-service Coverage Ratio	1.64	2.26	3.04
C	Average DSCR		2.31	
D	Return on Investment (ROI)	17.66	20.08	22.49
E	IRR		17%	

As perceived from the project cost and means of finance table, the suggested Debt-Equity Ratio for the proposed project is 2:1. The IRR for the proposed unit is approx. 17 per cent projected for a period of 10 years.

CLEARANCE REQUIRED

The proposed unit will have to register itself with Secretariat of Industrial Approvals, Ministry of Industries and Government of India, by filing Industrial Entrepreneur's Memorandum, as it will have plant and machinery investment of more than INR 10 million. The proposed unit is for manufacturing for food processing enzymes which are critical inputs for the consuming industries. There is fast growing demand of food processing enzymes in the global market and the project envisages export to advanced countries like USA, Canada, UK, Europe and Australia. The unit will require to register their products with Food and Drugs Administration in these countries, apart from registration with Indian and state food administration.

The most critical aspect of this product will be its shelf life for export consumers and hence it is critical to meet FDA regulations in consuming countries and Codex standards followed by them. The unit will get EOU registration from RBI, DGFT and with APEDA as registered manufacturer exporter to avail export incentives.

Being an exporting unit it will have to follow strict quality standards as accepted in the countries where export is to be made. Indian FDA authorities have laid down standards for food additives including food processing enzymes which will be followed by the proposed unit. It is obligatory to meet provisions under the PFA act for all ingredients and quality aspects for marketing the product in the Indian market.

FOOD MANUFACTURING/ PROCESSING OPERATIONS

The Food Code specifies under that the food establishment operator must obtain a variance from the regulatory authority for all food manufacturing/ processing operations based on the prior approval of a HACCP plan. The purpose of this is to provide processing criteria for different types of food manufacturing/processing operations for use by those preparing and reviewing HACCP plans and proposals. Criteria for additional processes will be provided as they are developed, reviewed, and accepted.

REDUCED OXYGEN PACKAGING (ROP)

ROP which provides an environment that contains little or no oxygen, offers unique advantages and opportunities for the food industry but also raises many microbiological concerns. Products packaged using ROP may be produced safely if proper controls are in effect. Producing and distributing these products with a HACCP approach offer an effective, rational, and systematic method for the assurance of food safety. The purpose of this Annex is to provide guidelines for effective food safety controls for retail food establishments covering the receipt, processing, packaging, holding, displaying, and labeling of food in reduced oxygen packages.

DEFINITIONS

The term ROP is defined as any packaging procedure that results in a reduced oxygen level in a sealed package.

The term is often used because it is an inclusive term and can include other packaging options such as:

- Cook-chill is a process that uses a plastic bag filled with hot cooked food from which air has been expelled and which is closed with a plastic or metal crimp.
- Controlled Atmosphere Packaging is an active system which continuously maintains the desired atmosphere within a package throughout the shelf-life of a product by the use of agents to bind or scavenge oxygen or a sachet containing compounds to emit a gas. Controlled Atmosphere Packaging is defined as packaging of a product in a modified atmosphere followed by maintaining subsequent control of that atmosphere.
- Modified Atmosphere Packaging is a process that employs a gas

flushing and sealing process or reduction of oxygen through respiration of vegetables or microbial action. Modified Atmosphere Packaging is defined as packaging of a product in an atmosphere which has had a one-time modification of gaseous composition so that it is different from that of air, which normally contains 78.08 per cent nitrogen, 20.96 per cent oxygen, 0.03 per cent carbon dioxide.

- Sous Vide is a specialized process of ROP for partially cooked ingredients alone or combined with raw foods that require refrigeration or frozen storage until the package is thoroughly heated immediately before service. The sous vide process is a pasteurization step that reduces bacterial load but is not sufficient to make the food shelf-stable. The process involves the following steps:
  - Preparation of the raw materials;
  - Packaging of the product, application of vacuum, and sealing of the package;
  - Pasteurization of the product for a specified and monitored time/ temperature;
  - Rapid and monitored cooling of the product at or below 3°C or frozen; and
  - Reheating of the packages to a specified temperature before opening and service.
- Vacuum Packaging reduces the amount of air from a package and hermetically seals the package so that a near-perfect vacuum remains inside. A common variation of the process is Vacuum Skin Packaging. A highly flexible plastic barrier is used by this technology that allows the package to mold itself to the contours of the food being packaged.

BENEFITS OF ROP

ROP can create a significantly anaerobic environment that prevents the growth of aerobic spoilage organisms, which generally are Gram negative bacteria such as Pseudomonas or aerobic yeast and moulds. These organisms are responsible for off-odours, slime, and texture changes, which are signs of spoilage. ROP can be used to prevent degradation or oxidative processes in food products. Reducing the oxygen in and around a food retards the amount of oxidative rancidity in fats and oils. ROP also prevents colour deterioration in raw meats caused by oxygen.

An additional effect of sealing food in ROP is the reduction of product shrinkage by preventing water loss. These benefits of ROP allow an extended shelf-life for foods in the distribution chain, providing additional time to reach new geographic markets or longer display at retail. Providing an extended shelf-life for ready-to-eat convenience foods and advertising foods as "Fresh-Never Frozen" are examples of economic and quality advantages.

SAFETY CONCERNS

Use of ROP with some foods can markedly increase safety concerns. Unless potentially hazardous foods are protected inherently, simply placing them in ROP without regard to microbial growth will increase the risk of foodborne illnesses.

ROP processors and regulators must assume that during distribution of foods or while they are held by retailers or consumers, refrigerated temperatures may not be consistently maintained. In fact, a serious concern is that the increased use of vacuum packaging at retail supermarket deli-type operations may be followed by temperature abuse in the establishment or by the consumer. Consequently, at least one barrier or multiple hurdles resulting in a barrier need to be incorporated into the production process for products packaged using ROP.

The incorporation of several sub-inhibitory barriers, none of which could individually inhibit microbial growth but which in combination provide a full barrier to growth, is necessary to ensure food safety. Some products in ROP contain no preservatives and frequently do not possess any intrinsic inhibitory barriers that either alone or in combination will inhibit microbial growth. Thus, product safety is not provided by natural or formulated characteristics. An anaerobic environment, usually created by ROP, provides the potential for growth of several important pathogens. Some of these are psychrotrophic and grow slowly at temperatures near the freezing point of foods.

Additionally, the inhibition of the spoilage bacteria is significant because without these competing organisms, tell-tale signs signaling that the product is no longer fit for consumption will not occur. The use of one form of ROP, vacuum packaging, is not new. Many food products have a long and safe history of being vacuum packaged in ROP. However, the early use of vacuum packaging for smoked fish had disastrous results, causing a long-standing moratorium on certain uses of this technology.

Refrigerated Holding Requirements for Foods in ROP

Safe use of ROP technology demands that adequate refrigeration be maintained during the entire shelf-life of potentially hazardous foods to ensure product safety. Bacteria, with the exception of those that can form spores, are eliminated by pasteurization. However, pathogens may survive in the final product if pasteurization is inadequate, poor quality raw materials or poor handling practices are used, or post-processing contamination occurs. Even if foods that are in ROP receive adequate thermal processing, a particular concern is present at retail when employees open manufactured products and repackage them. This operation presents the potential for post-processing contamination by pathogens.

If products in ROP are subjected to mild temperature abuse, i.e., 5°-12°C, at any stage during storage or distribution, foodborne pathogens, including *Bacillus cereus*, Salmonella spp., *Staphylococcus aureus*, and *Vibrio parahaemolyticus* can grow slowly. Marginal refrigeration that does not facilitate growth may still allow Salmonella spp., Campylobacter spp., and Brucella spp. to survive for long periods of time. Recent published surveys indicate that refrigeration practices at retail need improvement. Some refrigerated products offered in convenience stores were found at or above 7.2°C 50 per cent of the time; in several cases temperatures as high as 10°C were observed. Delicatessen display cases have been shown to demonstrate poor temperature control.

Foods have been observed above 10°C and above 12.8°C in several instances. Supermarket fresh meat cases appear to have a relatively good record of temperature control. However, even these foods can occasionally be found above 10°C. Temperature abuse is common throughout distribution and retail markets. Strict adherence to temperature control and shelf-life must be observed and documented by the establishment using ROP. Information on temperature control should also be provided to the consumer.

Currently these controls are not extensively used. Additionally, some commercial equipment is incapable of maintaining foods below 7.2°C because of refrigeration capacity, insufficient refrigerating medium, or poor maintenance. Most warehouses and transport vehicles in US distribution chains maintain temperatures in the 0°-3.3°C range. It must be assumed, however, for purposes of assessing risk, that occasionally temperatures of 10°C or higher may occur for extended periods. At retail, further temperature abuse must also be assumed. For instance, retail display cases can be as high as 13.3°C for short periods and some refrigerated foods are provided no refrigeration for short periods of time. These realities point to the need for establishments to implement controls, such as buyer specifications, over refrigerated distribution systems so that better temperature control can be ensured.

Control of Clostridium botulinum and *Listeria monocytogenes* in Reduced Oxygen Packaged Foods

Recently, there has been an increased interest in vacuum packaging or MAP at retail using conventional refrigeration for holding. Refrigerated foods packaged at retail may be chilled either after they are physically prepared and repackaged, or packaged after a cooking step. In either case but primarily the latter, germination of *Clostridium botulinum* spores must be inhibited because spores are not destroyed by a heating step. Sanitary safeguards must be employed to prevent reintroduction of pathogens.

Chief among these is *Listeria monocytogenes. Clostridum botulinum* is the causative agent of botulism, a severe food poisoning characterized by double vision, paralysis, and occasionally death. The organism is an anaerobic spore-forming bacteria that produces a potent neurotoxin. The spores are ubiquitous in nature, relatively heat-resistant, and can survive most minimal heat treatments that destroy vegetative cells. Certain strains of *C. botulinum*, which have been primarily associated with fish, are psychrotrophic and can grow and produce toxin at temperatures as low as 3.3°C. Other strains of *C. botulinum* can grow and produce toxin at temperatures slightly above 10°C. If present, *C. botulinum* could potentially grow and render toxigenic a food packaged and held in ROP because most other competing organisms are inhibited by ROP. Therefore, the food could be toxic yet appear organoleptically acceptable. This is particularly true of psychrotrophic strains of *C. botulinum* that do not produce tell-tale proteolytic enzymes. Because botulism is potentially deadly, foods held in anaerobic conditions merit regulatory concern and vigilance. The potential for botulism toxin to develop also exists when ROP is used after heat treatments such as pasteurization, or sous vide, processing of foods which will not destroy the spores of *C. botulinum*.

Mild heat treatments in combination with ROP may actually select for *C. botulinum* by killing off its competitors. If the applied heat treatment does not produce commercial sterility, the food requires refrigeration to prevent spoilage and ensure product safety. For this reason, sous vide products are frequently flash frozen in liquid nitrogen and held in frozen storage until use. There is a further microbial concern with ROP at retail. Processed products such as meats and cheeses which have undergone an adequate cooking step to kill *L. monocytogenes* can be contaminated when opened, sliced, and repackaged at retail. Thus, a simple packaging or repackaging operation can present an opportunity for recontamination with pathogens if strict sanitary safeguards are not in place. Processors of products using ROP should be cautious if they plan to rely on refrigeration as the sole barrier that ensures product safety. This approach requires very rigorous temperature controls and monitored refrigeration equipment. If extended shelf-life is sought, a temperature of 3.3°C or lower must be maintained at all times to prevent outgrowth of *C. botulinum* and the subsequent production of toxin. *Listeria monocytogenes* can grow at even lower temperatures; consequently, appropriate use-by dates must be established and readily apparent to the consumer. Since refrigeration alone does not guarantee safety from pathogenic microorganisms, additional growth barriers must be provided. Growth barriers are provided by hurdles such as low pH, aw, or short shelf life, and constant monitoring of the temperature. Any one hurdle, or a combination of several, may be used with refrigeration to control pathogenic outgrowth.

Design of Heat Processes for Foods in Reduced Oxygen Packages

Heat processes for sous vide or cook-chill operations should be designed so that, at a minimum, all vegetative pathogens are destroyed by a pasteurization process. Special labeling of these products is necessary to ensure adequate warning to consumers that these foods must be refrigerated at 5°C and consumed by the date required by the Code for that particular product. The National Advisory Committee on Microbiological Criteria for Foods chartered by the US Department of Agriculture and the Department of Health and Human Services recently commented on the microbial safety of refrigerated foods containing cooked, uncured meat or poultry products that are packaged for extended refrigerated shelflife and are ready-to-eat or prepared with little or no additional heat treatment.

The Committee recommended guidelines for evaluating the ability of thermal processes to inactivate *L. monocytogenes* in extended shelf-life refrigerated foods. Specifically, it recommended a proposed requirement for demonstrating that an ROP process provides a heat treatment sufficient to achieve a 4 decimal log reduction of *L. monocytogenes*. Other scientific reports recommend more extensive thermal processing. Thermal processes for sous vide practiced in Europe are designed to achieve a 12-13 log reduction of the target organism *Streptococcus faecalis*. It is reasoned that thermal inactivation of this organism would ensure destruction of all other vegetative pathogens.

Food manufacturers with adequate in-house research and development programmes may have the ability to design their own thermal processes. However, small retailers and supermarkets may not be able to perform the microbiological challenge studies necessary to provide the same level of food safety. If a retail establishment wishes to use an ROP process, microbiological studies should be performed by, or in conjunction with, an appropriate process authority or person knowledgeable in food microbiology who is acceptable to the regulatory authority. Finally, if foods are held long enough, even under proper refrigeration, extended shelf-life may be a problem. A recent study on fresh vegetables inoculated with *L. monocytogenes* was conducted to determine the effect of CAP on shelf life. The study found that CAP lengthened the time that all vegetables were considered acceptable, but that populations of *L. monocytogenes* increased during that extended storage.

Consumer Handling Practices and In-home Refrigeration Temperatures

Extended shelf-life provided by ROP is cause for concern because of the potential for abuse by the consumer. Consumers often can not, or do not, maintain adequate refrigeration of potentially hazardous foods at home. Foods in ROP that are taken home may not be eaten until enough time/temperature abuse has occurred to allow any pathogens present to increase to levels which can increase the chance of illness. Under the best of circumstances home refrigerators can be expected to range between 5° and 10°C. One study reported that home refrigerator temperatures in 21 per cent of the households surveyed were 10°C. Another study reported more than 1 of 4 home refrigerators are above 7.2°C and almost 1 of 10 are above 10°C. Thus, refrigeration alone cannot be relied on for ensuring microbiological safety after foods in ROP leave the establishment. Consumers have come to expect that certain packages of foods would be safe without refrigeration. Low-acid canned foods have been thermally processed, which renders the food shelf-stable. Retort heating ensures the destruction of *C. botulinum* spores as well as all other foodborne pathogens. Yet consumers may not understand that most products that are packaged in ROP are not commercially sterile or shelf-stable and must be refrigerated.

A clear label statement to keep the product refrigerated must be provided to consumers. The use of ROP has been extensively studied by regulators and the food industry over the past several years. Recommendations have been adapted from the Association of Food and Drug Officials "Retail Guidelines—Refrigerated Foods in Reduced Oxygen Packages" and New York State Department of Agriculture and Markets "Proposed Reduced Oxygen Packaging Regulations."

As provided in the Food Code, some ROP operations may be conducted under Reduced Oxygen Packaging, Criteria. Food that is packaged by an ROP method under these provisions is considered safe while it is under the control of the establishment and, if the labeling instructions are followed, while under the control of the consumer.

SAFETY BARRIER VERIFICATION

The safety barriers for all processed foods held in ROP at retail must be verified in writing. This can be accomplished through written certification from the product manufacturer. Independent laboratory analysis using methodology approved by the regulatory authority can also be used to verify incoming product and should be used to verify the barriers in a product that is packaged within the establishment by an ROP method. It should be noted that the Association of Food and Drug Officials guidelines recommend that laboratory analysis be conducted by official methods of the Association of Official Analytical Chemists. The multiple barrier or hurdle efficacy should be validated by inoculated pack or challenge studies. A product should be tested under abuse temperatures to demonstrate product safety during the food's shelf-life. Any changes in product formulation or processing procedures are cause for notification of the regulatory authority and a required approval of the revised ROP process. A record of all safety barrier verifications should be updated every 12 months. This record must be available to the regulatory authority for review at the time of inspection.

USDA PROCESS EXEMPTION

Meat and poultry products cured at a food processing plant regulated by the US Department of Agriculture using substances. Approval of substances for use in the preparation of products and Restrictions on the use of substances in poultry products are exempt from the safety barrier verification requirements.

PRINCIPLES OF THERMAL PROCESSING

The "canning" of foods has been practiced for almost 200 years, but the science supporting the canning process has been understood for only about half of that time. The theory and science that forms the basis for the development and application of thermal processes to low-acid and acidified foods packaged in hermetically sealed containers. The objectives of this section are for you to be able to:

- Define commercial sterility.
- Identify who can establish a thermal process.
- Identify the components in establishing a thermal process.
- Identify factors that impact the thermal process.
- Recognize a process deviation.

THE SCHEDULED PROCESS

As mentioned above, "canned" products are treated with heat to make them commercially sterile. The condition of commercial sterility is recognized as follows:

COMMERCIAL STERILITY

The condition achieved by application of heat, sufficient alone or in combination with other ingredients and/or treatments, to render the product free of microorganism capable of growing in the product at nonrefrigerated condition (over 50°F or 10°C) at which the product is intended to be held during distribution and storage. A condition of commercial sterility will result in products that are safe to eat because the pathogens of concern are destroyed or inactivated.

The product will remain shelf-stable as long as the container is intact because any spoilage organism that favours the environmental conditions within the container (i.e., anaerobic) and normal storage temperatures (i.e., mesophilic bacteria) are also destroyed with the thermal process. The current FSIS Canning Regulations refer to this condition as shelf-stability, but in this course we will refer to "canned" products as being commercially sterile to avoid confusion with the dry and semi-dry meat and poultry products that are also shelf-stable.

The application of heat to make a product commercially sterile is conducted in a controlled manner; this has been traditionally called the scheduled process or process schedule. The FSIS Canning Regulations require that scheduled processes be established by a processing authority. A *processing authority* is a person or organization having expert knowledge of thermal processing requirements for foods packed in hermetically sealed containers and having adequate facilities to make these process determinations.

The scheduled process, as designed by a processing authority, if properly executed, will produce a commercially sterile product. The scheduled process, includes thermal processing parametres such as initial temperature of the product, the process temperature and the process time, plus any other critical factors that may affect the attainment of commercial sterility. Critical factors may include any characteristic, condition or aspect of a product, including formulation, container, preparation procedures, or processing system that affect the scheduled process. Processing authorities use their knowledge of food microbiology (specifically, thermo bacteriology), heating characteristics of food, and processing systems as the basis for process schedule establishment. Contrary to other cooked meat and poultry products, the cook process for commercially sterile products is controlled by monitoring the process parameters (such as process temperature and time) rather than monitoring a temperature of the product. Temperature and time are easily controlled and monitored to ensure delivery of a proper thermal process.

For low-acid canned foods (those with a pH greater than 4.6), the thermal process focuses on the destruction of the spores of certain sporeforming bacteria. The target pathogen for low-acid canned foods is *Clostridium botulinum* (specifically the spores of the organism). Failure to destroy these spores, followed by germination and growth, can lead to the production of the deadly botulinum toxin, an extremely potent neurotoxin.

Low-acid canned food processes that assure the destruction of *C. botulinum* spores are adequate to protect human health; however, as noted previously, a more substantial process, or a commercial sterility process, is required to destroy spores of other microorganisms that could germinate and grow under normal storage and handling conditions and cause economic spoilage. The target for commercial sterility is typically *C. sporogenes*, an organism very similar to *C. botulinum* but with a higher heat resistance.

Thermal processes for acidified foods are targeted toward vegetative cells of microorganisms and are generally significantly milder than those applied to low-acid foods. This is primarily because spores of microorganisms such as *C. botulinum* will not germinate due to the acid nature in the product. Keeping the spores from germinating will prevent the growth of the vegetative cells of *C. botulinum* and subsequent toxin production. In distinguishing between acidified and low-acid foods, the standard used is a pH of 4.6 (greater than 4.6 is low-acid and less than or equal to 4.6 is acid or acidified).

Typical thermal processes for acidified foods will maintain the commercial sterility of a product as long as good sanitation and GMPs are followed. Lack of pH control for an acidified food can lead to problems if the pH is high enough to allow surviving spores of *C. botulinum* to germinate and produce toxin. Determining the scheduled process with the proper temperature and process time needed to produce commercially sterile products has been the subject of years and years of study in the canning industry. Sound process determinations depend upon good knowledge of the:

- Nature of the product and how it heats.
- Container in which the product is packed.

- Details of the thermal processing procedures used.
- Characteristics of the target microorganisms such as growth, survival, and thermal resistance.

These four factors are related to the thermal resistance of the microorganisms and the heating characteristics of the product. Utilizing all this information, the processing authority will establish a thermal process that will specify the amount of time at a specific temperature necessary to ensure the destruction of *C. botulinum* and spoilage organisms that may be present.

Establishing a Thermal Process

As mentioned previously, the processing authority will base the establishment of a thermal process for a particular food product on two separate factors:

- The thermal (also known as heat) resistance of the microorganism of choice in that food.
- The heating characteristics of the product.

In simplified terms, the processing authority needs to know:

- How much heat and for how long is necessary to destroy microorganisms in the food product.
- How fast does the product heat (in the case of conventional canning) or how does the product flow (in the case of aseptic processing).

The combination of these two factors is used to establish the thermal process. The establishment of the thermal process will also depend upon the method of processing: conventional canning or aseptic processing. In conventional canning, the product is filled into the container, the container is hermetically sealed, and the container and product are thermally processed at a specified time and temperature to achieve commercial sterility. For aseptic processing, packages or packaging material and the food product are sterilized in separate systems. Product sterilization involves heating a pumpable product to a sterilizing temperature and holding it at that temperature for sufficient time to sterilize the product. The packaging materials are sterilized with heat, chemicals, radiation or a combination. The sterile package is then filled with sterile product, closed and hermetically sealed in a sterile chamber. Once a process has been established for a particular food, it is specific for that particular set of parameters regarding formulation, preparation, thermal processing system, container, etc. Since a seemingly insignificant change in any of these parameters could result in under-processing, it is important that processes not be altered without consultation with a processing authority.

Thermal Resistance of Microorganisms

The thermal resistance of microorganisms (vegetative cells or spores) is dependent upon a number of factors:

- The growth characteristics of the microorganisms.

- The nature of the food in which the microorganisms are heated.
- The kind of food in which the heated microor-ganisms are allowed to grow.

Because of the variability of any biological entity, thermobacteriology is a highly complex science, and variations in any of these factors can affect the heat resistance of microorganisms.

Product Heating Data Determinations

Processing authorities determine product heating characteristics through the use of heat penetration tests. The rate at which a product heats can be measured using devices to monitor the rate of change in temperature of the food as it is being heated within the container (heat penetration studies). These determinations are made with a temperature sensor or thermocouple located in the product at the slowest heating region of the container.

The slowest heating region will depend on the type of product, container size, processing method and the heat transfer mechanism. Typical heat transfer mechanisms in canned foods are convection (heat current flowing within the container as experienced in canned broths), conduction (molecule to molecule as experienced in corned beef hash), and combinations (for example, convection then conduction due to thickening of the product as maybe experienced in formulated products such as chicken and dumplings).

The number of containers per run and the number of runs necessary to ensure the data are adequate depends on the variability of the product and the processing system. When either demonstrates significant variability, additional containers/runs may be required to ensure confidence in the process.

The need to simulate the worst case scenario likely to occur when producing product cannot be overstated, because the thermal process is controlled by monitoring and controlling the process parameters rather than with the actual temperature of the product.

The processing authority will review variables in product preparation such as changing a starch or protein in the formulation, filling procedures, rehydration procedures, etc., to determine the impact on the heating rate of the product.

More viscous product, higher fill weights, or improperly rehydrated product can all affect the heating rate of the product. If the heat penetration tests do not account for this effect or if the establishment can not control for these variations, the result could be under-processed product.

The heating characteristics of the product aseptically processed are also important. For aseptic processing, the heated product flows at a constant rate through tubing of a specified length; it is the flow characteristics of the product that determine whether certain particles in the fluid stream move through the tubing faster than others. The speed of the fastest moving particles (which remain at the elevated temperature for the shortest time) and the temperature of the flowing product determine the heat (lethality) attained by the product.

Acidified Foods

Products that are high acid (have a low pH) or are acidified to a pH of 4.6 or less do not require a high temperature process. The foods may be processed at the temperature of boiling water-212°F (100°C)-or lower. The thermal process is designed to destroy vegetative cells and some spores of low heat resistance. The product's low pH (4.6 or less) will prevent the remaining spores from growing out. To produce products with a pH of 4.6 or less, acidification must be properly carried out. Here are some methods to obtain properly acidified foods:

- Blanch the food ingredients in an acidified aqueous solution. To acidify large food particulates, the particulates could be blanched in a hot acid bath. The ability to obtain a properly acidified product is dependent upon blanch time and temperature, as well as the type of and concentration of acid.
- Immerse the blanched foods in an acid solution. That is, blanch the product in the normal steam or water blancher. Then, dip it into an acid solution, remove it from the acid solution and place it into containers. Proper acidification depends upon how well the product is blanched, the concentration of the acid and the contact time.
- Direct batch acidification is normally the best way to acidify fluid material. Ingredients are mixed in a kettle, and acid is added directly to the batch. (An elevated temperature may improve the rate of acid penetration into solid particles.) The pH of the batch is checked before the material is sent from the batch kettle to the filler.
- Add acid foods to low-acid foods in controlled portions. Essentially, this is how a formulated product such as pasta sauce is made. Components in the sauce, such as meat or onions, are low-acid foods, while the tomato sauce is an acid food. The acid food is mixed with the low-acid food to get an acidified food product. The formulation, including the proportion of tomato sauce to low-acid components, is critical to obtain uniform and accurate control of pH of the finished product.
- Directly add a predetermined amount of acid to individual containers during production. This involves addition of acid pellets, known volumes of acid solution, or some other means for direct acidification of each container. This is probably the most inaccurate and least consistent method of acidification, because acid addition to a given container may be overlooked. Although this is a permissible way to acidify, it is not recommended and it is not used for meat and poultry products.

Processes for acidified foods may involve a hot-fill-hold process or the conventional canning process. Hot-fill-hold processing involves filling the product hot, sealing the container, then holding filled containers for a period of time at or above a certain temperature before cooling. Acidified products may also be processed in a pasteurizer, atmospheric cooker or retort for a given period of time to destroy the target microorganisms. The processing authority will calculate a process using pH, fill temperature, and the thermal resistance data for the significant spoilage organisms.

Inoculated Test Pack

It is possible and sometimes desirable (e.g., with new products or new processing systems) to confirm the calculated process by means of inoculated test packs. In this procedure, the test product is prepared under commercial plant operating conditions. Appropriate test microorganisms of known heat resistance are added to the product. (It is undesirable to use *C. botulinum* spores in processing plants processing low-acid canned foods due to the risk associated with the potential toxin production.)

The product is inoculated with a known number of microorganisms and is then subjected to various processing times at one or a number of different processing temperatures. The product is then incubated at an appropriate temperature for growth of the test organism. Product that received a process inadequate to destroy the added microorganisms will show evidence of spoilage.

A satisfactory process is demonstrated by the absence of spoilage. Substantial agreement between calculated processes and those determined by inoculated packs furnishes the strongest possible assurance of the adequacy and safety of a particular process. Occasionally, due to peculiarities of the thermal processing systems or the product, reliable heating data cannot be obtained. Under these circumstances, consideration may be given to using an inoculated pack alone to establish a safe thermal process.

Process Deviations

A deviation in processing (or a process deviation) occurs whenever an actual thermal process is less than the scheduled process (or process schedule) or when any critical factor does not comply with the requirements for that factor as specified in the process schedule. Common causes of process deviations for traditional canning operations include failure to meet initial temperature, process time, or retort temperature requirements or failure to meet any other specified critical factor.

For aseptic processing and packaging systems, deviations can result from improper hold tube temperature, flow rate, or package sterilization; breach of the aseptic zone, etc. According to FSIS regulations deviations involving meat and poultry products may be handled in accordance with a HACCP plan for canned product that addresses hazards associated with microbial contamination or alternative documented proce-dures that will ensure that only safe and stable product is shipped in commerce. Product involved in deviations identified during processing may be immediately reprocessed using the full scheduled process, may be given an appropriate alternate process established in accordance with FSIS regulations, or may be held for later evaluation by a processing authority to determine its safety and stability.

It should be kept in mind that, in some cases, the original process may change the heating characteristics of the processed product (e.g., thickening due to starch). For this reason it is a good idea to consult with the processing authority to assure that the reprocess is adequate for the product and conditions of use. Upon completion of the evaluation of the process deviation, the establishment maintains a complete description of the deviation along with all necessary supporting documentation; a copy of the evaluation report; and a description of any product disposition actions, either taken or proposed.

If an alternate process schedule is used that is not on file with the inspector or if an alternate process schedule is immediately calculated and used, the product shall be set aside for further evaluation as noted above. Product involved in process deviations that are identified through review of processing and production records shall be held and the deviations evaluated by a processing authority in accord with the requirements noted above. If product involved in a deviation is destroyed, destruction shall be conducted in accordance with FSIS regulations. FSIS regulations also require the maintenance of a process deviation file.

The establishment shall maintain full records regarding the handling of each deviation, regardless of the seriousness of the deviation. Such records shall include, at a minimum, the appropriate processing and production records, a full description of the corrective actions taken, the evaluation procedures and results, and the disposition of the affected product. Such records shall be maintained in a separate file or in a log that contains the appropriate information. The file or log shall be retained for no less than one year at the establishment, and for an additional 2 years at a suitable location. The file or log shall be made available to inspection personnel upon request.

MARKET FOR HIGH-PRESSURE PROCESSED FOODS

Today the market for high-pressure processed foods is reaching 2 billion dollars annually. HPP applications include:

- *High-pressure Processing of Ready-to-eat Meat*: High-pressure Processing (HPP) provides quality extension and longer shelf life while reducing dependence on preservatives. Pathogens are inactivated by HPP and the level of spoilage organisms is reduced significantly. High pressure processed meats retain their original sensory qualities such as texture, colour and nutritional content throughout their shelf life. HPP can also more than double the shelf life of products using alternative preservation methods.

High pressure processing is a science-based post-lethality intervention step for ready-to-eat meats. HPP is not merely a surface treatment, but effective throughout the product package, whatever its size or shape. This is especially important to those meat processors producing sliced deli meats, where the risk of re-contamination with harmful pathogens, particularly *Listeria monocytogenes*, may be greatest. The Food Safety and Inspection Service published its Interim Final Rule in June 2003, which directs meat processors to declare the food safety protocols they use to control *L. monocytogenes*: High-pressure processing (HPP) gives consumers safe packaged fruit and vegetable products with fresh, just-prepared characteristics without additives or preservatives.

Fig. High Pressure Processed Foods.

The products add great consumer value by retaining the sensory qualities, texture, colour and nutritional content of the fresh-picked product with HPP, have almost unlimited opportunities to create exceptional value-added products. Many consumers were introduced to high-pressure processing with their first taste of all-natural refrigerated guacamole, soon followed by ripe avocado halves with up to 30-day shelf life. Avure's technology has expanded the universe of avocado products. This was only the beginning. Now, innovative food processors are merchandizing fresh, great-tasting products to supermarkets, club stores, food service distributors and major chain users. Salsa, pre-chopped onions, organic juices, flavourful fruit smoothies, colourful applesauce—these are just a few of the premium food items that only HPP makes possible.

High pressure processing inactivates *Salmonella, E. coli* and *Listeria monocytogenes* in fruit and vegetable products. Processors can use HPP in their HACCP programme to achieve the FDA requirement of 5-log reduction of pathogens in fresh juice. Consumers want foods that taste good and that are good for them. HPP makes this possible by retaining more of the taste, texture and vitamins that consumers want in fruits and vegetable products. High-pressure processing (HPP) leaves juice not only safe, but also with a fresh, just-squeezed appearance and flavour, natural texture and nutrition, without any need for additives or preservatives. The extended refrigerated shelf life achieved by HPP increases distribution opportunities, reduces returns and sensitivity to cool chain abuse, and allows more efficient production scheduling. A treatment at very high pressure, up to 600 MPa (87,000 psi) lasting less than a minute, results in the inactivation of spoilage organisms, including yeast and mould, and harmful pathogens, as well as the reduction of enzymatic activity. High pressure processing (HPP) gives manufacturers of ready-to-use wet salads and dips the tools to provide safer and more natural products with extended quality and shelf life. HPP offers natural, wholesome products without the need for heat or chemicals.

With the assistance of the high-pressure processing technology, some segments of the industry are also able to optimize organoleptical properties of their products because of the positive effects on food components such as hydrocolloids and proteins. The results are improved viscosity, mouth feel and the reduction of syneresis.

HIGH-PRESSURE FOOD PROCESSING OF RICE AND STARCH FOODS

The use of pressure (P) in addition to temperature (T) has been accepted in the field of food science and technology over the past 15 years, and research and development are now under way in the food industry, universities, and government institutes. Several commercial products are now on the market that are prepared using high-pressure techniques. This section describes the principle of high-pressure treatment and the effects of high pressure on foods, with emphasis on the high-pressure effects on starches. Finally, recent successes of the Echigo Seika Company in the application of high-pressure techniques to rice and rice products are discussed.

Principle and Method

High pressure means high hydrostatic pressure generated by the compression of water. A pressure of 100 MPa or higher (usually lower than 1,000 MPa) is used under temperatures below 100 °C. A unit of pressure is expressed in kg cm^{-2}, bar, or pounds in^{-2}, but now the international Pascal unit is commonly used. Therefore, 1,000 bar or 1,000 kg cm^{-2} almost equals 100 MPa. Food, which is contained in a plastic bag and sealed by carefully removing air, is placed in a pressure vessel filled with water and high pressure is then generated in the vessel by a pressure pump.

Pressurization of Food

An egg is not crushed by compression at 600 MPa, but the egg white and yolk are coagulated. The colour of egg yolk of a pressurized egg is naturally yellow, whereas a boiled egg changes to a faded yellow. The colour difference is attributed to changes associated with pressurization: pressure affects only noncovalent bonding and coagulates proteins without splitting covalent bonds, thus keeping the colour and smell intact. High pressure induces protein denaturation in the same way as high temperature. Meat protein is also denatured by pressure treatment at 400 MPa, preserving the properties of raw meats. An example of prawns or shrimp is interesting: although a boiled prawn turns red and the meat coagulates, the appearance of a pressurized prawn is the same as that of raw shrimp, but the meat coagulates after pressurization at 400 MPa for 10 min.

High pressure also has an effect on starches: a thick suspension of rice starch forms a ball by standing against its own weight after pressurization at 700 MPa, indicating that starches are gelatinized by the pressure treatment.

Versatile Utility of High Pressure

A high-pressure treatment generally coagulates protein, thereby inactivating the enzymes, gelatinizing the starches, and killing microorganisms. Thus, the use of high pressure promises to be a versatile process in food science and technology. Pressure produces a new texture in meats and starch-based foods, while keeping the original nutrients, flavour, and colour.

High-Pressure Effects on Pure Starches

High-pressure-induced Gelatinization of Starch

Effect of Pressure on Amylase Digestibility of Starches. The starches of potato, maize, and wheat are gelatinized by pressure treatment at warm conditions of 45–50°C. The pressurization produces unique properties that are different from those of heat-gelatinization: heat-treatment destroys starch granules, resulting in a transparent solution, but a pressure-treatment swells the granules while maintaining the granular structure. Nevertheless, amylolytic enzymes such as α-, β-, and glc-amylases digest the pressurized starches well, being similar to the phenomenon in which the pressure-treatment of proteins increases protease digestibility. The pressure-induced gelatinization of starches exhibits a sigmoid curve, suggesting that a two-state transition is involved, as in heat-induced gelatinization.

Effect of Pressurization Time on Amylase Digestibility of Starches. To attain full amylase digestibility of starches, pressurization under warm conditions for 2 to 6 h is necessary. Interestingly, pressurization of starches for a longer time makes amylase digestion difficult: the amylase digestibility of starches decreased by 20–50 per cent after pressurization for 17 h compared with the maximum digestibility obtained after pressurization for 2 to 6 h.

These observations suggest that pressure induces gelatinization of starches, similar to heating, but prolonged pressurization produces a new stable structure of starches, which is not susceptible to attack by amylase. *Birefringence of Starches after Pressurization*. The birefringence of starches is lost as increasing pressure is applied. Wheat starch is sensitive to pressure and birefringence is lost at 200 MPa. In the pressurization of starches, the number of granules exhibiting complete birefringence decreases without increasing the incomplete birefringent granules. This loss of birefringence shows that the crystalline structure is destroyed by high pressure as well as by high temperature and that the high-pressure-induced gelatinization of starches follows a two-state transition without an intermediate state of destruction.

Physical Properties of Pressurized Starches

When a 50 per cent water suspension of starches is pressurized at 100 to 500 MPa and at 45°C for 1 h and air-dried, followed by analyses of their physical properties, the results are as follows. According to X-ray crystal analysis, the crystalline structure of potato, waxy maize, and maize starches decreases with an increase in pressure, although the crystalline structure of potato starch does not change up to 500 MPa. The decrease in the high-pressure-induced crystalline structure is parallel with the increase in amylase susceptibility. Amylograms show that the transition temperature of pressurized starches is elevated, thus decreasing the viscosity. Interestingly, differential-scanning calorimetry (DSC) of pressurized starches shows that the peak temperature of the DSC patterns increases while the peak area decreases, indicating that some pressure-induced structural state of the pressurized starches is perturbed by low energy at higher temperatures.

Summary of Pressure-induced Changes in Starches

The structure of pressurized starches changes accompany-ing the increase in amylase digestibility, a loss of birefringence, and a loss of crystallinity. These unique structural properties, which are somewhat different from those of heat-gelatinized starches, should be analysed in detail for effective applications of high pressure to starch and related foods.

Rice-based Foods Produced by High-pressure Processing

Dr. Akira Yamazaki, of Echigo Seika Company Ltd., studied the properties of pressurized rice in detail to introduce the use of high pressure to the food industry. He equipped several large highpressure machines in his own factory and succeeded in sending rice and rice products to the market by introducing the high-pressure technique to the manufacturing process of rice products, cooked rice (*gohan*), rice crackers (*osembei*), and rice cakes (*omochi*).

High-pressure Effects on Rice Grains

Traditionally cooked-rice grains change their shape and develop some cracks, but rice grains after high-pressure pretreatment followed by heat-cooking swell, and their original shape is maintained without cracks.

High-pressure Cooked Rice for Microwave Ovens

Bread is purchased in the store and toasted just before eating. This is a typical life style, especially on a busy morning. However, 45 min are required to cook rice. Cooked rice tastes best and has its best texture just after steaming. Warming of cold cooked rice makes the taste worse. In general, heating a starch-based food twice leads to an unpalatable food. In history, instant-rice is a dream of Japanese consumers.

Dr. Yamazaki succeeded in producing oven-cooked rice with a good taste and texture for consumers, although the production scale is only enough to fulfill the requirement of a small city. He also succeeded in producing instant rice containing miscellaneous cereals. These instant cooked-rice cereals exhibited good taste and good texture by a 3-min heating in a microwave oven.

When high-pressure pretreated and successively heatcooked rice is compared with traditionally cooked rice, its properties are as follows: first, it is more gelatinized; second, it is more slowly retrograded; and third, it is gelatinized to a greater extent by heating just before serving. Japanese consumers who are particularly sensitive to rice accept these properties.

High-pressure Rice Cakes

During New Year days, the Japanese public is freed from the kitchen and enjoys the New Year cerebrations, eating rice cakes (*omochi*) every day, which are a preserved food. The company introduced the high-pressure technique for producing the traditional rice cake and a special kind of *omochi*, which contains herbs with a natural colour, smell, and taste, and this is now available on the market.

Merits of High-pressure Food Processing

High-pressure processing is useful not only for producing high-quality food, but also for improving the manufacturing process. The production time for rice crackers is shortened by introducing a high-pressure technique into the traditional processing system and the total energy cost is decreased by 10 per cent and the labour force by 56 per cent. The use of high pressure for food processing and cooking, in addition to heating and cooling, is now in our hands. Two factors, T and P, are useful for the manufacturing of good foods, including starch-based foods. Although T and P are used independently, their combined use is also important for the optimal use of high pressure. For example, heat-tolerant bacterial spores are inactivated by pressurization under elevated temperature.

FRUIT PRESERVES

Fruit preserves are fruits, or vegetables, that have been prepared and canned or sealed air-tight for long term storage. The preparation of fruit preserves traditionally involves the use of pectin as a gelling agent, although sugar or honey may be used as well. The ingredients used and how they are prepared will determine the type of preserves; jams, jellies and marmalades are all examples of different styles of fruit preserves that vary based upon the ingredients used. There are various varieties of fruit preserves made globally, and they can be made from sweet or savory ingredients. In North America, the plural form preserves is used, while the singular preserve is used in British and Commonwealth English. Additionally, the name of the type of fruit preserves will also vary depending on the regional variant of English being used.

VARIATIONS

Chutney

A chutney is a pungent relish of Indian origin made of fruit, spices and herbs. Although originally intended to be eaten soon after production, modern chutneys are often made to be sold and so require a preservatives—often sugar and vinegar—to ensure it has a suitable shelf life. Mango chutney, for example, is mangoes reduced with sugar.

Confit

Confit, which is the past participle form of the French verb "confire" or "to preserve", is most often applied to preservation of meats, especially poultry and pork, by cooking them in their own fat or oils and allowing the fats to set. However, the term can also refer to fruit or vegetables which have been seasoned and cooked with honey or sugar until it has reached a jam-like consistency. Savory confits, such as ones made with garlic or tomatoes, may call for a savory oil such as virgin olive oil as the preserving agent.

Conserves

A conserve is a jam made of fruit stewed in sugar. Often the making of conserves can be trickier than making a standard jam, because the balance between cooking, or sometimes steeping in the hot sugar mixture for just enough time to allow the flavour to be extracted from the fruit, and sugar to penetrate the fruit, and cooking too long that fruit will break down and liquefy. This process can also be achieved by spreading the dry sugar over raw fruit in layers, and leaving for several hours to steep into the fruit then just heating the resulting mixture only to bring to the setting point. As a result of this minimal cooking, some fruits are not particularly suitable for making into conserves, because they require cooking for longer periods to avoid issues such as tough skins. Currants and gooseberries, and a number of plums are among these fruits. Due to this shorter cooking period, not as much pectin will be released from the fruit, and as such, conserves will sometimes be slightly softer set than some jams. An alternate definition holds that conserves are preserves made from a mixture of fruits and/or vegetables. Conserves may also include dried fruit or nuts.

Fruit Butter

Fruit butter, in this context, refers to a process where the whole fruit is forced through a sieve or blended after the heating process:

- "Fruit butters are generally made from larger fruits, such as apples, plums, peaches or grapes. Cook until softened and run through a sieve to give a smooth consistency. After sieving, cook the pulp... add sugar and cook as rapidly as possible with constant stirring.... The finished product should mound up when dropped from a spoon, but should not cut like jelly. Neither should there be any free liquid."

Fruit Curd

Fruit curd is a dessert topping and spread usually made with lemon, lime, orange, or raspberry. The basic ingredients are beaten egg yolks, sugar, fruit juice and zest which are gently cooked together until thick and then allowed to cool, forming a soft, smooth, intensely flavoured spread. Some recipes also include egg whites and/or butter.

Fruit Spread

Fruit spread refers to a jam or preserve with no added sugar.

Jam

Jam contains both fruit juice and pieces of the fruit's flesh, however, some cookbooks define jam as cooked and gelled fruit purees. Properly, the term jam refers to a product made with whole fruit, cut into pieces or crushed. The fruit is heated with water and sugar to activate the pectin in the fruit. The mixture is then put into containers. The following extract from a US cookbook describes the process:

- "Jams are usually made from pulp and juice of one fruit, rather than a combination of several fruits. Berries and other small fruits are most frequently used, though larger fruits such as apricots, peaches, or plums cut into small pieces or crushed are also used for jams. Good jam has a soft even consistency without distinct pieces of fruit, a bright colour, a good fruit flavour and a semi-jellied texture that is easy to spread but has no free liquid."

Jelly

Jelly is a clear or translucent fruit spread made from sweetened fruit juice and set using naturally occurring pectin. Additional pectin may be added where the original fruit does not supply enough, for example with grapes. Jelly can be made from sweet, savory or hot ingredients. It is made by a process similar to that used for making jam, with the additional step of filtering out the fruit pulp after the initial heating. A muslin or stockinette "jelly bag" is traditionally used as a filter, suspended by string over a bowl to allow the straining to occur gently under gravity. It is important not to attempt to force the straining process, for example by squeezing the mass of fruit in the muslin, or the clarity of the resulting jelly will be compromised.

- "Good jelly is clear and sparkling and has a fresh flavour of the fruit from which it is made. It is tender enough to quiver when moved, but holds angles when cut.
- *Extracting Juice*: Pectin is best extracted from the fruit by heat, therefore cook the fruit until soft before straining to obtain the juice.... Pour cooked fruit into a jelly bag which has been wrung out of cold water. Hang up and let drain. When dripping has ceased the bag may be squeezed to remove remaining juice, but this may cause cloudy jelly."

Marmalade

British-style marmalade is a sweet preserve with a bitter tang made from fruit, sugar, water, and a gelling agent. American-style marmalade is sweet, not bitter. In English-speaking usage, "marmalade" almost always refers to a preserve derived from a citrus fruit, most commonly oranges, although onion marmalade is also used as an accompaniment to savoury dishes. The recipe includes sliced or chopped fruit peel, which is simmered in fruit juice and water until soft; indeed marmalade is sometimes described as jam with fruit peel. Such marmalade is most often consumed on toasted bread for breakfast.

The favoured citrus fruit for marmalade production in the UK is the "Seville orange", *Citrus aurantium* var. aurantium, thus called because it was originally imported from Seville in Spain; it is higher in pectin than sweet oranges, and therefore gives a good set. Marmalade can also be made from lemons, limes, grapefruit, strawberries or a combination.

REGIONAL TERMINOLOGY

The term preserves is usually interchangeable with jam. Some cookbooks define preserves as cooked and gelled whole fruit which includes a significant portion of the fruit.

The terms jam and jelly are used in different parts of the English-speaking world in different ways. In the United States, both jam and jelly are sometimes popularly referred to as "jelly", whereas in the United Kingdom, Canada, India and Australia, the two terms are more strictly differentiated. In Australia and South Africa, the term jam is more popularly used as a generic term for both jam and jelly. To further confuse the issue, the term jelly is also used in the UK, South Africa, Australia, India and New Zealand to refer to a gelatin dessert, known in North America as jello, derived from the brand name Jell-O.

PRODUCTION

In general, jam is produced by taking mashed or chopped fruit or vegetable pulp and boiling it with sugar and water. The proportion of sugar and fruit varies just as to the type of fruit and its ripeness, but a rough starting point is equal weights of each. When the mixture reaches a temperature of 104°C, the acid and the pectin in the fruit react with the sugar, and the jam will set on cooling. However, most cooks work by trial and error, bringing the mixture to a "fast rolling boil", watching to see if the seething mass changes texture, and dropping small samples on a plate to see if they run or set. Commercially produced jams are usually produced using one of two methods. The first is the open pan method, which is essentially a larger scale version of the method a home jam maker would use. This gives a traditional flavour, with some caramelization of the sugars.

The second commercial process involves the use of a vacuum vessel, where the jam is placed under a vacuum, which has the effect of reducing its boiling temperature to anywhere between 65-80°C depending on the recipe and the end result desired. The lower boiling temperature enables the water to be driven off as it would be when using the traditional open pan method, but with the added benefit of retaining more of the volatile flavour compounds from the fruit, preventing caramelization of the sugars, and of course reducing the overall energy required to make the product. However, once the desired amount of water has been driven off, the jam still needs to be heated briefly to 95-100 °C to kill off any micro-organisms that may be present; the vacuum pan method does not kill them all.

During the commercial filling of the jam into jars, it is common to use a flame to sterilize the rim of the jar and the lid to destroy any yeasts and molds which may cause spoilage during storage. It is also common practice to inject steam into the head space at the top of the jar immediately prior to the fitting of the lid, in order to create a vacuum. Not only does this vacuum help prevent the growth of spoilage organisms, it also pulls down the tamper evident safety button when lids of this type are employed. How easily a jam sets depends on the pectin content of the fruit.

Some fruits, such as gooseberries, redcurrants, blackcurrants, most citrus fruits, apples, and raspberries, set very well; others, such as strawberries and ripe blackberries, often need to have pectin added. There are commercial pectin products on the market, and most industrially-produced jams use them. Home jam-makers sometimes rely on adding a pectin-rich fruit to a poor setter, for example apple to blackberries. Other tricks include extracting juice from lemons, redcurrants or gooseberries, or making a pectin stock with whole apples or just the cores and skins; once cooled, this "stock" can then be frozen for later use. Making jam at home is a popular handicraft activity, and many take part in this. Homemade jam may be made for personal consumption, or as part of a cottage industry.

PRINCIPLES OF FOOD PRESERVATION

You have learnt earlier that by boiling milk we are preserving it for a longer time. But, what are you actually doing by boiling? You are killing the microorganisms by raising the temperature of milk. Microorganisms cannot survive at very high temperature. This is one of the principles of food preservation. Let us now learn about the principles of food preservation:

- By killing the microorganisms.
- By preventing or delaying the action of micro-organisms.
- By stopping the action of enzymes.

BY KILLING THE MICRO-ORGANISMS

You already know that boiling of milk kills micro-organisms. Sometimes, heat is applied for a shorter duration to kill only undesirable microorganisms, that is those which can spoil the food stuff. It is done while pasteurizing milk. The cooking that you do at home also keeps food free from microorganisms. In canning, food is heated to high temperature to prevent growth of microorganisms in food.

PREVENTING OR DELAYING THE ACTION OF MICROORGANISMS

You all know that a peeled apple spoils faster than one with intact skin. Do you know why? This is because the apple has its skin as a protective covering which prevents the entry of microorganisms. Similarly, the shell of nuts and eggs, skin of fruits and vegetables serve as a protective coating and delays the action of microorganisms. Food packed in polythene bags and aluminium foils are also protected against microorganisms. You have read earlier that microorganisms need air and water to grow. But if these are removed, you can prevent the action of microorganisms and ensure that food does not get spoilt.

Lowering temperature or freezing a food also helps in delaying the action of microorganisms and thus in food preservation. You must have come across frozen foods. Frozen food can be kept for a longer time than fresh food. This is because microorganisms cannot act at low temperatures. Thus, when you are putting food in the refrigerator or freezer, you are preventing the microorganism from growing. Lastly, certain chemicals like sodium benzoate and potassium metabisulphite also help in preventing the growth of micro-organisms. These chemicals are called "preservatives".

Thus you have learnt that the action of microorganisms can be delayed or prevented in many ways:

- By providing a protective covering.
- By raising the temperature.
- By lowering the temperature.
- By adding chemicals.

BY STOPPING THE ACTION OF ENZYMES

Enzymes also cause food spoilage. They are naturally present in food. Take the example of fruits. Keep a raw banana for a few days and observe what happens to it. Yes, the banana will turn ripe, become yellow and then start decaying and browning.

All this happens due to presence of enzymes. What will happen if the action of enzymes is stopped? The foodstuff will be prevented from spoiling. Enzyme action can be prevented by giving a mild heat treatment. Before canning or freezing, vegetables are dipped in hot water or exposed to steam for a few minutes. This is known as blanching. When you heat milk, you are not only killing microorganisms present in it but also stopping the action of enzymes. This extends its shelf life.

HOUSEHOLD METHODS OF FOOD PRESERVATION

Some of the practical methods which can be used for preserving food at home are:

- Dehydration.
- Pickling with salt, spices and/or oil.
- Making jams, jellies, murabbas.
- Bottling of squashes and juices.
- Freezing.

CANNING

Canning is a method of preserving food in which the food is processed and sealed in an airtight container. The process was first developed as a French military discovery by Nicolas Appert in 1810. The packaging prevents microorganisms from entering and proliferating inside. To prevent the food from being spoiled before and during containment, quite a number of methods are used: pasteurization, boiling, refrigeration, freezing, drying, vacuum treatment, antimicrobial agents that are natural to the recipe of the foods being preserved, a sufficient dose of ionizing radiation, submersion in a strong saline solution, acid, base, osmotically extreme or other microbe-challenging environments.

Other than sterilization, no method is perfectly dependable as a preservative. For example, the microorganism *Clostridium botulinum* can only be eliminated at temperatures above the boiling point.

From a public safety point of view, foods with low acidity need sterilization under high temperature. To achieve temperatures above the boiling point requires the use of a pressure canner. Foods that must be pressure canned include most vegetables, meats, seafood, poultry, and dairy products. The only foods that may be safely canned in an ordinary boiling water bath are highly acidic ones with a pH below 4.6, such as fruits, pickled vegetables, or other foods to which acidic additives have been added.

HISTORY AND DEVELOPMENT OF CANNING

During the first years of the Napoleonic Wars, the French government offered a hefty cash award of 12,000 francs to any inventor who could devise a cheap and effective method of preserving large amounts of food. The larger armies of the period required increased, regular supplies of quality food. Limited food availability was among the factors limiting military campaigns to the summer and autumn months. In 1809, a French confectioner and brewer, Nicolas Appert, observed that food cooked inside a jar did not spoil unless the seals leaked, and developed a method of sealing food in glass jars. The reason for lack of spoilage was unknown at the time, since it would be another 50 years before Louis Pasteur demonstrated the role of microbes in food spoilage. However, glass containers presented challenges for transportation. Glass jars were largely replaced in commercial canneries with cylindrical tin or wrought-iron canisters following the work of Peter Durand. Cans are cheaper and quicker to make, and much less fragile than glass jars. Glass jars have remained popular for some high-value products and in home canning. Can-openers were not invented for another thirty years—at first, soldiers had to cut the cans open with bayonets or smash them open with rocks. The French Army began experimenting with issuing canned foods to its soldiers, but the slow process of canning foods and the even slower development and transport stages prevented the army from shipping large amounts across the French Empire, and the war ended before the process was perfected.

Unfortunately for Appert, the factory which he had built with his prize money was razed in 1814 by Allied soldiers invading France. Following the end of the Napoleonic Wars, the canning process was gradually employed in other European countries and in the US. Based on Appert's methods of food preservation, Peter Durand patented a process in the United Kingdom in 1810. He did not develop the process, selling his patent in 1811 to Bryan Donkin and John Hall, who were in business as Donkin Hall and Gamble, of Bermondsey. Bryan Donkin developed the process of packaging food in sealed airtight cans, made of tinned wrought iron. Initially, the canning process was slow and labour-intensive, as each large can had to be hand-made, and took up to six hours to cook, making canned food too expensive for ordinary people. The main market for the food at this stage was the Army and Navy. By 1817 Donkin recorded that he had sold £3000 worth of canned meat in six months. In 1824 Sir William Edward Parry took canned beef and pea soup with him on his voyage to the Arctic in HMS Fury, during his search for a northwestern passage to India. In 1829, Admiral Sir James Ross also took canned food to the Arctic, as did Sir John Franklin in 1845.

Some of his stores were found by the search expedition led by Captain Leopold McLintock in 1857. One of these cans was opened in 1939, and was edible and nutritious, though it was not analyzed for contamination by the lead solder used in its manufacture. Throughout the mid-19th century, canned food became a status symbol amongst middle-class households in Europe, becoming something of a frivolous novelty. Early methods of manufacture employed poisonous lead solder for sealing the cans, which may have worsened the disastrous outcome of the 1845 Franklin expedition to chart and navigate the Northwest Passage. Increasing mechanisation of the canning process, coupled with a huge increase in urban populations across Europe, resulted in a rising demand for canned food. A number of inventions and improvements followed, and by the 1860s smaller machine-made steel cans were possible, and the time to cook food in sealed cans had been reduced from around six hours to thirty minutes. Canned food also began to spread beyond Europe—Robert Ayars established the first American canning factory in New York City in 1812, using improved tin-plated wrought-iron cans for preserving oysters, meats, fruits and vegetables.

Demand for canned food greatly increased during wars. Large-scale wars in the nineteenth century, such as the Crimean War, American Civil War, and Franco-Prussian War introduced increasing numbers of working-class men to canned food, and allowed canning companies to expand their businesses to meet military demands for non-perishable food, allowing companies to manufacture in bulk and sell to wider civilian markets after wars ended. Urban populations in Victorian era Britain demanded ever-increasing quantities of cheap, varied, quality food that they could keep at home without having to go shopping daily. In response, companies such as Nestlé, Heinz, and others emerged to provide quality canned food for sale to working class city-dwellers. In particular, Crosse and Blackwell took over the concern of Donkin Hall and Gamble. The late 19th century saw the range of canned food available to urban populations greatly increase, as canners competed with each other using novel foodstuffs, highly decorated printed labels, and lower prices.

Demand for canned food skyrocketed during World War I, as military commanders sought vast quantities of cheap, high-calorie food to feed their millions of soldiers, which could be transported safely, survive trench conditions, and not spoil in transport. Throughout the war, soldiers generally subsisted on low-quality canned foodstuffs, such as the British "Bully Beef", pork and beans and Maconochies Irish Stew, but by 1916 widespread boredom with cheap canned food amongst soldiers resulted in militaries purchasing better-quality food to improve morale, and the complete meals in a can began to appear. In 1917 the French Army began issuing canned French cuisine, such as coq au vin, while the Italian Army experimented with canned ravioli and spaghetti bolognese. Shortages of canned food in the British Army in 1917 led to the government issuing cigarettes and amphetamines to soldiers to suppress their appetites. After the war, companies that had supplied military canned food improved the quality of their goods for civilian sale. Today, tin-coated steel is the material most commonly used. Laminate vacuum pouches are also used for canning, such as used in MREs and Capri Sun drinks.

DOUBLE SEAMS

Invented in 1888 by Max Ams, modern double seams provide an airtight seal to the tin can. This airtight nature is crucial to keeping bacteria out of the can and keeping its contents sealed inside. Thus, double seamed cans are also known as Sanitary Cans. Developed in 1900 in Europe, this sort of can was made of the traditional cylindrical body made with tin plate. The two ends were attached using what is now called a double seam. A can thus sealed is impervious to the contamination by creating two tight continuous folds between the can's cylindrical body and the lids.

This eliminated the need for solder and allowed improvements in manufacturing speed, reducing cost. Double seaming uses rollers to shape the can, lid and the final double seam. To make a sanitary can and lid suitable for double seaming, manufacture begins with a sheet of coated tin plate. To create the can body, rectangles are cut and curled around a die, and welded together creating a cylinder with a side seam. Rollers are then used to flare out one or both ends of the cylinder to create a quarter circle flange around the circumference. Precision is required to ensure that the welded sides are perfectly aligned, as any misalignment will cause inconsistent flange shape, compromising its integrity. A circle is then cut from the sheet using a die cutter. The circle is shaped in a stamping press to create a downward countersink to fit snugly in to the can body. The result can be compared to an upside down and very flat top hat. The outer edge is then curled down and around about 140 degrees using rollers to create the end curl. The result is a steel tube with a flanged edge, and a countersunk steel disc with a curled edge. A rubber compound is put inside the curl.

Seaming

The body and end are brought together in a seamer and held in place by the base plate and chuck, respectively. The base plate provides a sure footing for the can body during the seaming operation and the chuck fits snugly in to the end. The result is the countersink of the end sits inside the top of the can body just below the flange. The end curl protrudes slightly beyond the flange.

First Operation

Once brought together in the seamer, the seaming head presses a first operation roller against the end curl. The end curl is pressed against the flange curling it in towards the body and under the flange. The flange is also bent downward, and the end and body are now loosely joined together. The first operation roller is then retracted. At this point five thicknesses of steel exist in the seam.

From the outside in they are:

- End,
- Flange,

- End Curl,
- Body,
- Countersink.

This is the first seam. All the parts of the seam are now aligned and ready for the final stage.

Second Operation

The seaming head then engages the second operation roller against the partly formed seam. The second operation presses all five steel components together tightly to form the final seal.

The five layers in the final seam are then called:

1. End,
2. Body Hook,
3. Cover Hook,
4. Body,
5. Countersink.

All sanitary cans require a filling medium within the seam because otherwise the metal-to-metal contact will not maintain a hermetic seal. In most cases, a rubberized compound is placed inside the end curl radius, forming the critical seal between the end and the body.

Probably the most important innovation since the introduction of double seams is the welded side seam. Prior to the welded side seam, the can body was folded and/or soldered together, leaving a relatively thick side seam. The thick side seam required that the side seam end juncture at the end curl to have more metal to curl around before closing in behind the Body Hook or flange, with a greater opportunity for error.

Seamer Setup and Quality Assurance

Many different parts during the seaming process are critical in ensuring that a can is airtight and vacuum sealed. The dangers of a can that is not hermetically sealed are contamination by foreign objects or that the can could leak or spoil. One important part is the seamer setup. This process is usually performed by an experienced technician. Amongst the parts that need setup are seamer rolls and chucks which have to be set in their exact position.

The lifter pressure and position, roll and chuck designs, tooling wear, and bearing wear all contribute to a good double seam. Incorrect setups can be non-intuitive. For example, due to the springback effect, a seam can appear loose, when in reality it was closed too tight and has opened up like a spring. For this reason, experienced operators and good seamer setup are critical to ensure that double seams are properly closed. Quality control usually involves taking full cans from the line—one per seamer head, at least once or twice per shift, and performing a teardown operation, mechanical tests as well as cutting the seam open with a twin blade saw and measuring with a double seam inspection system. The combination of these measurements will determine the seam's quality. Use of a or software in conjunction with a manual double seam monitor, computerized double seam scanner, or even a fully-automatic double seam inspection system makes the laborious process of double seam inspection faster and much more accurate.

Statistically tracking the performance of each head or seaming station of the allows for better prediction of can seamer issues, and may be used to plan maintenance when convenient: rather that to simply react after bad or unsafe cans have been produced.

Nutrition Value

Canning is a way of processing food to extend its shelf life. The idea is to make food available and edible long after the processing time. A 1997 study found that canned fruits and vegetables provide as much dietary fibre and vitamins as the same corresponding fresh or frozen foods, and in some cases, even more. The heating process during canning appears to make dietary fibre more soluble, and therefore more readily fermented in the colon into gases and physiologically active byproducts. Canned tomatoes have a higher available lycopene content.

POTENTIAL HAZARDS

To improve food safety for those who eat canned food, governments have enacted laws requiring alphanumeric codes being put on food cans during manufacture indicating information relevant to health, such as the date of canning, etc.

Migration of Can Components

In canning toxicology, migration is the movement of substances from the can itself into the contents. Potential toxic substances that can migrate are lead, causing lead poisoning, or bisphenol A, a potential endocrine disruptor that is an ingredient in the epoxy commonly used to coat the inner surface of cans.

Salt Content

Canned food can be a major source of dietary salt. Too much salt increases the risk of health problems, including high blood pressure. Therefore, health authorities have recommended limitations of dietary sodium. Many canned products are available in low-salt and no-salt alternatives.

Botulism

Foodborne botulism results from contaminated foodstuffs in which *C. botulinum* spores have been allowed to germinate and produce botulism toxin, and this typically occurs in canned non-acidic food substances. *C. botulinum* prefers low oxygen environments, and can therefore grow in canned foods. Botulism is a rare but serious paralytic illness, leading to paralysis that typically starts with the muscles of the face and then spreads towards the limbs. In severe forms, it leads to paralysis of the breathing muscles and causes respiratory failure. In view of this life-threatening complication, all suspected cases of botulism are treated as medical emergencies, and public health officials are usually involved to prevent further cases from the same source.

CANNING AND THE RECESSION

Canned goods sell particularly well in times of recession due to the tendency of financially-stressed individuals to engage in cocooning, a term used by retail analysts to describe the phenomenon in which people actively avoid straying from their houses. In February 2009, the recession-laden United States saw an 11.5 per cent rise in sales of canning-related items.

HYGIENE CO OL MEASURES IN FOOD PROCESSING

Hygiene in food processing started with the introduction of general measures, including cleaning and disinfection, prevention of re-contamination and treatment of food products to kill any microbial pathogens present. Heat treatment was introduced into food processing even before the underlying causes of foodborne illness were known. It was Nicholas Appert in France and Peter Durand in England who introduced canning of food and the use of thermal processing around 1800.

However, neither Appert nor Durand understood why thermally processed foods did not spoil and remained safe to eat. Then, Louis Pasteur showed that certain bacteria were either associated with food spoilage or caused specific diseases.

Based on Pasteur's findings, commercial heat treatment of wine was first used in 1867 to destroy any undesirable microorganisms, and the process was described as "pasteurization". This process was also recommended by Escherich to decontaminate milk.

In the course of time, it became clear that the effects of certain antimicrobial treatments were predictable. Two historical examples were the setting of performance criteria for destroying spores of *Clostridium botulinum* in low-acid, canned foods by Esty and Meyer and the process criteria for *Coxiëlla burnetii* in milk pasteurization, as determined by Enright *et al.* Further research resulted in predictions relating to many other processes, such as acidification, drying and the use of curing agents in meat products, on both pathogenic and spoilage organisms.

Such knowledge ushered in a new era in safe food production. This era is characterized by the division of hygiene measures into specific practices that are controllable and other general measures, the effects of which are largely unpredictable at present.

General Hygiene Practices

One of the first safety systems developed by the food industry was that involving the application of Good Manufacturing Practices (GMP), as a supplement to end-product testing. GMP covers all aspects of production, from starting materials, premises and equipment to the training of staff, and the WHO has established detailed guidelines.

GMP also provide a framework for hygienic food production, which is often referred to as Good Hygienic Practice (GHP). The establishment of GHP is the outcome of long practical experience and major components of the system are:

- *Design of Premises and Equipment*: This includes the location and layout of the premises to avoid hygiene hazards and facilitate safe food production. Food processing and handling equipment should always be designed with hygiene in mind, including ease of cleaning.
- *Control of the Production Process*: Control measures are applied throughout the supply chain and cover factors such as raw materials, packaging and process water, as well as the product itself. Key aspects include management and supervision of the process as a whole, as well as appropriate recording systems.
- *Plant Maintenance and Cleaning*: Both processing equipment and the fabric of the building should be maintained in good order. Suitable programmes need to be developed for plant cleaning and disinfection, and their effectiveness monitored routinely. Systems are also needed for pest control and management of waste.
- *Personal Hygiene*: Staff are required to maintain high standards of personal hygiene in relation to wearing of protective clothing, hand washing and general behaviour. Visitors must also be strictly controlled in these respects. The health status of personnel should be monitored regularly and any illness or injuries recorded.
- *Transportation*: Requirements should be established for the use and maintenance of transport vehicles, including their cleaning and disinfection. Vehicle usage should be managed and supervised.
- *Product Information and Consumer Awareness*: It is important that the final product is suitably labelled and that the consumer is provided with all relevant information on product handling and storage, including a "use-by" date. Labelling should also indicate the batch and origin of the product, so that full traceability is possible.
- *Staff Training*: In relation to food hygiene and safety, all personnel should receive appropriate training and be made fully aware of their individual responsibilities. Such training should be repeated and updated as required.

The GHP concept is largely subjective and its benefits tend to be qualitative rather than quantitative. It has no direct relationship to the safety status of the product, but its application is considered to be a necessary preventive measure in producing safe food.

Those hygiene measures that have a predictable outcome and are subject to control can be incorporated in the Hazard Analysis Critical Control Point (HACCP) concept. This concept seeks, among other things, to avoid reliance on microbiological testing of the end-product as a means of controlling food safety. Such testing may fail to distinguish between safe and unsafe batches of food and is both time-consuming and relatively costly. However, effective application of the HACCP concept depends upon GHP being used.

ENVIRONMENTAL ISSUES IN FOOD PROCESSING AND PACKAGING

Just about all leading food companies in Western Europe have, or are in the process of implementing, an internal environmental programme, since environmental effects have become a very important issue with consumers and government agencies alike. Consent to use "green" symbols on the food package is important not only to company image and sales but often also to processing efficiency and product cost. A powerful tool in the continuing work in these programmes is the use of Life Cycle Analysis (LCA), covering the interaction with the environment, including energy utilization, over the entire life cycle of food products, packaging, process machinery, etc.

It has, for example, become very important in identifying critical points in the food chain with regard to the environment and in choice situations between different alternatives, for example, for the packaging of a new product, or the choice between several alternative processes.

While food waste is equivalent to some economical loss and can mean short range pollution of the environment, even if composted or used as animal feed, it is 100 per cent biodegradable and a renewable resource. The situation for food packaging is different, and demands are getting stricter all the time from government agencies and consumers for reuse, recirculation, minimizing the use of packaging and "overpackaging" and designing packaging for easy separation of materials for reuse.

All packaging on the EU market should be manufactured according to the European Parliament and Council Directive 94/62/EC on packaging and packaging waste. The demands on packaging are the following:

- Preventing the production of packaging waste.
- Reusing packaging.
- Recycling and other forms of recovering packaging waste.
- The safe final disposal of packaging waste.
- Minimize the amount of packaging.
- Reduce landfill to the absolute minimum.

Reuse and Recycling

Within the EU research programmes a great amount of joint research has been devoted to problems in reuse and recycling of packaging, of importance to consumers safety and product quality. The reuse of PET bottles for beverages is very high, but cleaning has been problematic because water temperatures above 70°C cannot be used.

Flavour carry over can be considerable, since the commercial type of washing operation does not remove more than 20-50 per cent of sorbed volatiles from the previously packed product. Misuse by consumers, storing non-food liquids in the bottles prior to recirculation, is another problem, because of which standardized inertness and contamination tests have been developed.

Applying a protective layer of new PET on recycled PET proved to be an effective measure against carry over of volatiles. It was also demonstrated, that PET and PC bottles showed no thermal or chemical degradation and no change in barrier properties after 15 use cycles. Lately, there has been fairly strong pressure from governments and ecologist groups to recycle packaging back into food contact use, rather than for odd uses or incineration. One interesting project of this kind is presently being commercialized in Sweden in cooperation with Dow Chemical. Polyester or paperboard waste is mixed with food grade clay, latex and carbon black, formed into food serving trays and dried to physical stability and strength in a 200 kW microwave plant.

Edible and Biodegradable Packaging

There is a marked interest in Europe today in edible and biodegradable packaging and coating of foods and a great many research groups are involved in this area of packaging. As is well known, Japan already has developed a number of commercial edible film materials, both carbohydrate- and protein-based. In Europe, where research interest seems to be focused on protein based and starch derivative-based films, there is so far very little commercial application as film materials.

Fig. Smart Pack Hot Beverage System

Their excellent oxygen barrier properties first have to be combined with good machineability and mechanical strength and acceptable moisture barrier properties. Preliminary results indicate that applying a SiOx coating on edible films may lead to a solution. The European market leader in fresh dairy products, Danone, recently decided to use yoghurt cups made of polylactic acid, which is fully degradable, after development work in cooperation with Dow Chemical and Cargill.

Walki Lamicoat in co-operation with British Zeneca have developed a biodegradable packaging material consisting of paper or board laminated with the ICI Biopol material, which is made from sugar by bacterial fermentation. Wile environmental aspects are becoming increasingly important, convenience and novelty factors will sometimes be at least equally important. A good example is the coffee preparation package and server, now under introduction in Sweden. A rectangular paperboard package contains a closed PE capsule of instant coffee, surrounded by pure spring water. It is heated in a specially designed microwave oven. Correct final temperature is assured within 1°C by a liquid crystal tag, which changes colour. An optical sensor reacts on the colour change to shut off microwave power. At this temperature the capsule releases and mixes the coffee powder into the hot water, and fresh coffee can be served through the convenient pouring device.

8

Food Biotechnology

INTRODUCTION AND OBJECTIVES

Current and potential future benefits of biotechnology are numerous. They include new and more effective pharmaceuticals, more accurate methods of diagnosing disease, improved waste management techniques, new methods of production and improved food quality and safety. For people in Asia – whether they are wheat farmers on the Ganges Plain of India, rice farmers on the terraces of Bali, cotton growers in China or inhabitants of fast-paced cities like Singapore or Tokyo – the implications of biotechnology will be significant.

In the area of food and agriculture, the effects of biotechnology are likely to have many facets including better crop yields, reduced use of chemicals for plant disease and pest control, less environmental degradation, as well as the development of innovative food products, such as foods with improved nutritional value or better food quality and safety. Biotechnology is not seen as a panacea for all the region's food production problems. The supply of food is a complex process and biotechnology offers one part of a multi-faceted strategy to meet the growing demands for more and better quality foods. The Asian Food Information Centre and the International Service for the Acquisition of Agribiotech Applications have joined together to develop Food Biotechnology: A Communications Guide to Enhance Understanding Biotechnology. The Guide is intended to provide leaders in the scientific, medical, food and agricultural communities and educators involved in these areas, with a helpful resource kit. This resource is intended to provide context and to improve understanding of the practice and issues associated with modern biotechnology.

The kit aims to provide the most scientifically sound and up-to-date information about biotechnology products and processes. The resource provides an overview of regional production and regulations regarding crop biotechnology and includes findings from research conducted in Asian countries on consumer understanding and awareness of biotechnology. A presentation outlining the basic science of biotechnology is included in PowerPoint format with an accompanying commentary.

This could be used as a draft for presentations on Food Biotechnology or as a starting point for presentations in other languages. Future resources are planned which will provide information relevant to other areas of modern agricultural biotechnology. Given the diversity of communities in Asia with respect to race, religion, and social and economic status, it is especially important that any communication be based on an appreciation and respect for one's audience. It is also important to establish a baseline for the level of the understanding of food biotechnology present in your audience.

AGRICULTURAL BIOTECHNOLOGY AND FOOD SECURITY

Can molecular biology-based research contribute to the solution of the problems outlined earlier? Are the potential social and economic benefits likely to exceed potential risks or costs? If these questions are answered in the affirmative, issues related to the design of the technology and the needed policies and institutions must be tackled.

Although conventional applications of biotechnology, such as tissue culture and fermentation amongst others, is under way in several developing countries, little genetically improved seed material has been grown in the poorer developing countries to date so ex post assessment is virtually impossible. A great deal is known, however, about the social and economic risks and benefits associated with traditional Mendelian plant breeding as exemplified by the Green Revolution.

The analysis, therefore, begins with the identification of similarities and differences between the Green Revolution and modern biotechnology, and an attempt is made to draw sessions from the Green Revolution and to look at the difference between that technology package and modern biotechnology to try to assess the likely social and economic risks and benefits of modern agricultural biotechnology ex ante.

COMPARING THE GREEN REVOLUTION AND MODERN BIOTECHNOLOGY

Shift to Private Sector Research

There are three differences of particular importance for an assessment of social and economic risks and benefits. The research leading to the Green Revolution was undertaken by the public sector and the improved seed was usually freely available for seed multiplication and distribution. Although breeders' rights may permit an initial charge for the improved materials, the intellectual property rights did not extend beyond the initial release.

Having acquired the seed, farmers could reuse it without further payment, although reuse of hybrid seed would drastically reduce the yield advantage. This is in keeping with the principle of "farmers' rights" included in the 1983 International Undertaking on Plant Genetic Resources.

In contrast, the bulk of modern agricultural biotech-nology research is undertaken by private sector firms, which protect IPRs through patents that extend beyond the first release.

Farmers, therefore, cannot legally plant or sell for planting the crop produced with the patented seed without the permission of the patent holder. Patent holders, currently seeking ways to enforce their rights, are considering approaches such as legal agreements and technologies that will activate and deactivate specific genes. However, monitoring and enforcing contracts that prohibit large numbers of small farmers from using the crops they produce as seed would be expensive and difficult. The so-called terminator gene is the first patented technology aimed at biological IPR protection. It is not appropriate for small farmers in developing countries because existing infrastructure and production processes may not be able to keep fertile and infertile seeds apart.

Small farmers could face severe consequences if they planted infertile seeds by mistake. Commercialization of the terminator gene now seems unlikely in the short term. Research is under way on other biological approaches to IPR protection that would not impose such risk on small farmers. These include, for example, genetically engineered seeds that contain desired traits, such as pest resistance or drought tolerance, but in which these are activated only through chemical treatment. Otherwise, the seed would maintain its normal characteristics. Thus, if a farmer planted an improved seed, the offspring would not be sterile; rather they would revert to normal seeds, without the improved traits. The farmer would have the choice of planting the seed and doing no more, or activating the improved traits by applying the chemical. This approach complies with the principle of doing no harm.

It is important to note that even when patents permit a private company to enjoy monopoly or near-monopoly rights over a product it has developed, the firm is unlikely to capture 100 per cent of the economic benefits. A recent study of the distribution of the economic benefits generated by the use of herbicide-tolerant soybean seed in the United States in 1997 found that the company, Monsanto, received 22 per cent, while seed companies gained 9 per cent. Consumers of soybean and soybean products in the United States and other countries reaped a 21 per cent share, whereas farmers worldwide obtained 48 per cent. The share of U.S. farmers was actually 51 per cent of the benefits, but farmers elsewhere experienced net losses of 3 per cent.

Rise of Proprietary Research Processes and Technologies

A second, and related, difference between the Green and Gene revolutions involves the patenting of processes as well as products. The main process behind the Green Revolution was conventional plant breeding technology, which lies in the public domain, carried out by public institutions. Today, the processes used in modern agricultural biotechnology are increasingly subjected to IPR protection, along with the products that result. This means that public sector research institutions may not be able to gain access to basic but proprietary knowledge and processes needed in research, including research on the so-called orphan crops such as cassava and millet.

These are critical staples in the diets of many poor people, but they do not offer promising economic returns to private sector R&D efforts, so efforts to develop disease-resistant cassava or drought-tolerant millet, whether through genetic modification or conventional breeding, must come from the public sector. Some firms have agreed to transfer proprietary technologies, without charging royalties, to developing countries where there are few potential commercial prospects. Monsanto, for example, has entered into agreements with Kenyan and Mexican government agricultural research institutes to develop virus-resistant crops. Arrangements such as these are few and generally involve the philanthropic arms of the private firms.

Enlightened Adaptation Vs. Direct Transfer

A third difference involves the adaptation of industrial country agricultural research to developing country conditions. Although based on earlier research in industrial countries, the Green Revolution was focused on solving specific problems in developing countries. Current application of modern biotechnology is focused on industrial country agriculture. Industrial country research institutions had begun working on development of higher yielding crop varieties in the late 19th century. For example, in Japan, rice breeding under the auspices of the Ministry of Agriculture and public universities led to large yield gains in the early part of the 20th century, with a second wave of major gains after 1945. During the early decades of Soviet history, under the leadership of geneticist Nikolai Ivanovich Vavilov, the government carried out extensive crop improvement programmes and established one of the world's largest germplasm collections. In the United States, hybrid maize research began in the 1920s.

Much of the basic research was done by public institutions, such as land grant universities, state experiment stations, and the U.S. Department of Agriculture. Applications to particular farming conditions and the mass marketing of the new varieties were, in turn, handled by private seed firms such as Pioneer Hi-Bred and DeKalb. The research focused not only on developing higher yielding seeds to bolster food supplies for domestic consumption but also on animal feed and production for export. This research could not simply be transferred to poorer developing countries, where the need was for improved varieties of locally-consumed staples.

The research that led to the Green Revolution involved further adaptation to the agroecological conditions of tropical and semitropical areas. It also focused on rice, wheat, maize, root and tuber crops, and tropical fruits and vegetables. The public sector role was, if anything, even more prominent, with international agricultural research centres and national agricultural research systems, particularly in Asia and Latin America, playing a prominent role. Financial support came from donors of official development assistance and large private foundations, such as Ford, Rockefeller, and Kellogg.

In contrast, modern agricultural biotechnology is still in an early phase, and the focus is overwhelmingly on production on industrial country farms and for industrial country markets. In 1998, 85 per cent of the land planted to genetically improved crops was in just five developed countries with the United States alone accounting for about 75 per cent of the area. Argentina, China, Mexico, and South Africa cultivated the remaining 15 per cent, and the countries other than China include a substantial number of large-scale, capital-intensive farms that produce primarily for industrial country markets. Among the crops produced in these four developing countries are insect-resistant cotton and maize, herbicide-resistant soybean, and tomatoes with a long shelf life. Globally, herbicide- resistant soybean, insect-resistant maize, and genetically improved cotton account for 85 per cent of all plantings.

Both the area planted to genetically improved crops and the value of the harvests grew dramatically between 1995 and 1999: from less than 1 million hectares to 28 million in 1998 and approximately 40 million in 1999, and from US$75 million in 1995 to US$1.64 billion in 1998. Private industry has dominated research. Consolidation of the industry has proceeded rapidly since 1996, with more than 25 major acquisitions and alliances worth US$15 billion. Little private-sector agricultural biotechnology research so far has focused on developing country food crops other than maize. Moreover, little adaptation of the research to developing country crops and conditions has occurred through the "enlightened" public and philanthropic channels prominent in the Green Revolution of the developing countries.

Some of the exciting international and regional programmes are described by Cohen. A programme directed at public/private sector linkages is that of the International Service for the Acquisition of Agri-biotech Applications which transfers and delivers appropriate biotechnology applications to developing countries and builds partnerships amongst institutions. Relatively little biotechnology research currently focuses on the productivity and nutrition of poor people. The Rockefeller Foundation's agriculture programme is one example; in 1998, it provided about US$7.4 million for biotechnology research relevant to developing countries, mainly through IARCs and NARS in developing countries, with a major emphasis on rice. This sum pales by comparison with Monsanto's 1998 R&D budget of US$1.3 billion, much of which funded agricultural biotechnology research. As with the Green Revolution, the challenge is to move from the scientific foundation established by industrial country-oriented research efforts to research focused on the needs of poor farmers and consumers in developing countries. Direct transfers of the fruits of agricultural biotechnology research to the developing countries will not work, in most cases. More appropriate research for the developing world might focus on biotechnology and conventional breeding to develop alternative forms of weed resistance, such as leafier rice that denies weeds sunlight rather than incorporating herbicide tolerance into rice.

The West Africa Rice Development Association, a public IARC in Côte d'Ivoire, has used a combination of conventional plant breeding and tissue culture to develop such rice. Insect-resistant crops would have great potential value for poor farmers. So far, however, the development of crops containing genes from the Bacillus thuringiensis bacterium, which produces a natural pesticide, has focused largely on the crops and cropping environments of North America. The new crop varieties containing the Bt gene require extremely knowledge-intensive cultivation. They might well be transferable to larger scale operations in some developing countries such as Argentina. The potential usefulness of this application in crops grown by small farmers is open to question.

There is considerable debate about risks of the development of resistance in pests, harm to beneficial insects, and crosspollination of wild and weedy plants with the novel gene. The evidence on these issues is still inconclusive and warrants careful monitoring before the application of Bt is tried on a large scale in crops grown by subsistence farmers. Research on crops and problems of relevance to small farmers in developing countries will require the allocation of additional public resources to agricultural research, including biotechnology research, that promises large social benefits. There is no reason to believe that this research will offer lower rates of return than other agricultural research and development. Private-sector agricultural research currently accounts for a small share of agricultural research in most developing countries. The public sector can expand private-sector research for poor people by converting some of the social benefits to private gains, for example, by offering to buy exclusive rights to newly developed technology and make it available either for free or for a nominal charge to small farmers. The private research agency would bear the risks, as it does when developing technology for the market.

IARCs have an important role to play as intermediaries in facilitating such arrangements. Without more enlightened adaptation, continued expansion of genetically improved crop production in the industrial countries may well have a negative impact on small farmers in developing countries. Some developing country consumers would benefit, but those consumers who also farm could experience net losses. In addition, the development of industrial substitutes for developing country export crops, such as cocoa could have a devastating impact on developing country farmers' livelihoods. In sum, the biggest risk of modern biotechnology for developing countries is that technological development will bypass poor farmers and poor consumers because of a lack of enlightened adaptation.

It is not that biotechnology is irrelevant, but that research needs to focus on the problems of small farmers and poor consumers in developing countries. Private sector research is unlikely to take on such a focus, given the lack of future profits. Without a stronger public sector role, a form of "scientific apartheid" may well develop, in which cutting edge science becomes oriented exclusively towards industrial countries and large-scale farming.

Sessions from the Green Revolution

The outcomes of the Green Revolution offer some guideposts for assessing the likely risks and benefits of agricultural biotechnology for developing countries. Risks and benefits may be inherent in a given technology, or they may transcend the technology. The policy environment into which a technology is introduced is critical. For example, IFPRI research has found that in Tamil Nadu State in India, the adoption of high-yielding grain varieties meant not only increased yields and cheaper, more abundant food for consumers, but income gains for small and largerscale farmers alike, as well as for nonfarm poor rural households. Increased rural incomes contributed to nutrition gains for these households.

Because the Tamil Nadu state government has pursued active poverty alleviation strategies, including extensive social safety net programmes and investment in agriculture, rural development, and a fair measure of equity in access to resources such as land and credit, the benefits were widely shared. Where increased inequality followed the adoption of Green Revolution technology, it was not because of factors inherent to the technology, but rather a result of policies that did not promote equitable access to resources. And even in these areas, rural landless laborers usually found new job opportunities as a consequence of increased agricultural productivity, particularly where appropriate physical infrastructure and markets developed. Successful adoption of Green Revolution technology, however, depended on access to water, fertilizer, and pesticides.

Thus, inequality between well-endowed and resource-poor areas increased because of the properties of the technology itself. Likewise, excessive or improper use of chemical inputs led to adverse environmental impacts in some instances. This problem was offset, to some extent, by characteristics that were also inherent in the technology: by allowing yield gains without expanding cultivated area, the technology kept cultivators from clearing forests and moving onto wild and marginal lands. Overall, the Green Revolution was extremely successful in enhancing productivity in rice, wheat and maize; in increasing incomes and reducing poverty; and in preserving forests and marginal lands by improving yields within existing cultivated areas. By reducing unit costs and prices for food, it greatly benefited poor consumers, and by boosting farmers' incomes, it contributed to gains in nutrition.

Would agricultural biotechnology produce similar results in developing countries? The answer depends on whether the research is relevant to poor people and on its ownership, that is, the nature of the intellectual property rights arrangements.

AGRICULTURAL BIOTECHNOLOGY AND FOOD SAFETY

When a consumer buys food s/he expects it to be safe and have the same quality characteristics that s/he is accustomed to, and familiar with. Consumers have the right to know that this is the case and if there is anything different about that product. The Food and Agriculture Organization and the World Health Organization of the United Nations advocate the concept of 'substantial equivalence' as the most practical approach to address the safety evaluation of foods or food components derived through modern biotechnology. This approach states that 'if a new food or food component is found to be substantially equivalent to an existing food or food component, it can be treated in the same manner with respect to safety.' So what kind of tests are used to determine if the food is as safe or safer than existing foodstuffs? Researchers must prepare comprehensive data to support the safety and wholesomeness of new crop varieties developed through biotechnology.

This process requires years of laboratory and field testing before a product can be brought to the market. To provide assurance that foods derived through biotechnology are as safe as those produced by traditional breeding programmes, the safety assessment strategies involve several key steps. These steps include molecular characterisation of the genetic modification, agronomic characterisation, nutritional assessment, toxicological assessment and safety assessment.

For example, typical questions that must be addressed are:

- Does the genetically modified food have a traditional counterpart that has a history of safe use?
- Has the concentration of any naturally occurring toxins or allergens in the food changed?
- Have the levels of key nutrients changed?
- Do new substances in the genetically modified food have a history of safe use?
- Has the food's digestibility been affected?
- Has the food been produced using accepted, established procedures?

Even after these and other questions about the biotechnology derived foods are answered, there are still more steps in the approval process before the crop can be commercialized. In fact, genetically modified foods are the most studied food products ever produced.

- *Safety assessment*: Molecular characterisation – for new plant varieties produced through modern biotechnology, the source of the gene introduced into the plant is first identified. The transformation system used to insert the gene is defined, as well as the number of copies of inserted genes and the integrity and stability of the genetic insert.

- *Agronomic traits*: Usually the starting points for evaluating substantial equivalence. For example, in the case of potatoes, the traits commonly examined are yield, tuber size and distribution, dry matter content and disease resistance.
- *Nutritional assessment*: Involves key nutrients including fats, proteins, carbohydrates and essential vitamins and minerals.
- *Toxicology assessment*: Toxicants and anti-nutrients are compounds known to be naturally present in some crops that could have an impact on health if their levels increased. For example, solanine glycoalkaloids in potatoes or trypsin inhibitors in soybeans). The levels of anti-nutrients in crops produced through biotechnology are compared to conventionally produced varieties grown under comparable environmental and agronomic conditions.
- *Safety assessment*: When a crop produced through biotechnology is shown to be substantially equivalent to a conventional crop, the safety assessment focuses on the introduced trait and the expressed protein product.

The overall goal of these tests is to determine whether the plant is substantially equivalent to food derived from a conventional source that has a history of safe use. A substantial equivalence evaluation focuses on the product rather than the process used to develop the product. If the new product is substantially equivalent to the conventional food or feed, then the product derived through biotechnology is considered to be as safe as the conventional counterpart. If the food produced using biotechnology contains a new trait, which changes the levels of nutrients or antinutrients, such as a higher level of a vitamin or a lower level of an allergen, the assessment focuses on demonstrating the safety of the new trait.

ALLERGENS

One of the public's biggest concerns related to genetically modified foods is that an allergen could be accidentally introduced into a food product. The DNA inserted into a plant is safe and no case of allergy to DNA is known, the concern is whether the protein produced by the plant as a result of the inserted DNA could cause an unexpected allergic reaction. Scientists know a lot about which foods trigger allergic reactions in adults and children. Ninety per cent of all food allergies are associated with only eight foods or food groups – shellfish, eggs, fish, milk, peanuts, soybeans, tree nuts, and wheat. These, and many other food allergens, are well characterized. To date, genes found to express any known allergenic protein have been specifically excluded from the research and development process.

Scientists have agreed that following these simple steps provides compelling assurance that no allergic effects will result from foods derived from biotechnology:

- Avoid transferring genes from foods known to be allergenic.

- Check the structure of any new proteins produced in foods derived from biotechnology against the structures of known allergens to assure that no allergenic structures exist in the new protein.
- Measure the stability of the new protein in stomach and intestinal fluids. Most allergens are stable to these conditions. Proteins which are unstable to these conditions are not likely to be allergens.
- Determine how much of the new protein will be present in the food consumed by humans. Most allergens are present in large amounts.

Allergenicity screening continues to be a very important part of safety testing before a crop can enter into the food market.

ANTIBIOTIC RESISTANCE

Some biotechnology crops contain genes for a trait called antibiotic resistance. Scientists use this trait as a marker to identify cells into which the desired gene has been successfully introduced. Concerns have been raised that these marker genes could move from biotechnology crops to microorganisms that normally reside in a person's gut and lead to an increase in antibiotic resistance.

There have been numerous scientific reviews and experimental studies of this issue and they have come to the following conclusions:

- The likelihood of antibiotic resistance genes moving from biotechnology crops to other organisms is extremely remote.
- In the unlikely event that an antibiotic resistance gene is transferred to another organism, the impact of this transfer would be negligible, as the markers used in genetically modified crops have limited or no clinical or veterinary use.

However, in response to public concerns, scientists have been advised to avoid using antibiotic resistance genes in biotechnology modified plants. Alternative marker strategies are being developed.

FOOD SAFETY CAPACITY DEVELOPMENT

In order to support countries wishing to fulfil the mandate set by the CPB, the secretariat of the CBD maintains a global database of capacity-building initiatives as a component of the Biosafety Clearing House. The purpose of this database is to give an overview of past, present and prospective capacity-building initiatives. The secretariat intends to use the information to develop a method for coordinating capacity-building initiatives, thus ensuring that they complement one another, use funding efficiently and strengthen resources in the recipient countries. Although the secretariat's interest is in initiatives that would support the effective implementation of the CPB, the database covers a wider range of initiatives, such as technology transfer and those directed towards biotechnology research. To date, 89 initiatives have been listed in the database, illustrating a broad range of implementing agencies. The secretariat, more than half of the registered initiatives have been negotiated bilaterally and through industry interest groups. United Nations agencies, intergovernmental organizations, or individual governments, industry or NGOs supported most of these countries through bilateral agreements. While the capacity-building initiatives collectively cover all the aspects associated with the application of modern biotechnology, no single one covers the entire range—each is limited to its own specific focus. For example, the FAO/WHO expert consultations and capacity-building programmes supported by both organizations train individuals in food-safety-related issues only.

WHO has advised Member States and assisted in building their capacity for food-safety-related issues for many years. The food-safety activities of WHO have increased significantly over the years, with the establishment of international expert scientific bodies such as the Joint FAO/WHO Expert Committee on Food Additives in 1956, to evaluate the safety of food additives, contaminants, naturally occurring toxicants and residues of veterinary drugs in food; the Joint FAO/WHO Meeting on Pesticide Residues to evaluate the safety of pesticide residues in food; the Joint FAO/WHO Expert Meeting on Microbiological Risk Assessment to provide risk-assessment guidelines for selected pathogens and for microbiological hazards in food and water. Also in 1963, the Codex Alimentarius Commission was created to implement the joint FAO/WHO Food Standards Programme. To strengthen its in-house activities, WHO created the Food Safety Programme in 1978, operating at national, regional and international levels. The recognition of food safety as a major public-health concern by the World Health Assembly in 2000 has also increased the profile of food-safety-related issues, not only within the Organization but also at the national level.

These activities were further supported by the endorsement of the WHO Global Strategy for Food Safety by the WHO Executive Board in 2002. In this strategy, WHO proposes to "formulate regional food safety strategies on the basis of the WHO global food safety strategy and of specific regional needs such as technical support, educational tools and training". Considerable technical assistance has been provided to developing countries to create and/or enhance food-safety control systems, but these activities have not been effectively coordinated and therefore not been adequate in meeting the public-health needs of recipient countries.

The SPS Agreement of WTO calls for assistance to developing-country Members to enable them to strengthen their food safety and animal and plant health protection. The Agreement encourages Members to enter into bilateral arrangements for technical assistance, or to seek training through other international organizations. Such assistance can be in the area of processing technology, research or infrastructure development, and may take the form of technical advice, expertise, financial assistance or procurement of adequate equipment. Food-safety activities within WHO take place at the international, regional and country levels. The regional and country offices provide assistance in developing and strengthening national food-safety programmes, whereas WHO headquarters develops guidelines for such work, including the framework for risk analyses and setting international standards. The division of these activities is arbitrary as headquarters also participates in activities at the national and regional levels, with technical know-how and capacity-building guidance.

These activities include:

- Developing regional and national food-safety policy and strategies;
- Preparing food legislation, regulations, standards and codes of hygienic practice;
- Implementing food inspection programmes;
- Promoting methods and technologies designed to prevent foodborne diseases, including the hazard analysis and critical control point system;
- Developing or enhancing food analysis capability;
- Developing methods for assessing the safety of the products of new technologies;
- Establishing healthy markets and enhancing the safety of street food; and
- Promoting the establishment of foodborne disease surveillance systems.

Many WHO activities to build food-safety capacity are developed in collaboration with FAO. However, FAO also administers a major, separate technical cooperation programme building capacity in this and other agriculture-related areas in many developing countries. Although most developing countries have national food-control systems, these are often not based on modern scientific concepts. Moreover, they cannot be adapted to cope with developments in food science and technology. The specifications for an effective food-control system include: regulations, capacity for assessing the risks associated with the food, and ongoing monitoring and evaluation of the risks.

A capacity-building programme for the risk assessment of products of modern biotechnology would involve:

- Use of the concept of a comparative safety assessment;
- Hazard identification and characterization;
- Assessment of food intake, including consumption profile and effects;
- Use of integrated toxicological evaluation;
- Use of integrated nutritional evaluation;
- Risk characterization; and
- Application of risk-manageent strategies, such as labelling and monitoring.

Other Considerations

competent authorities need information relating to trends in biotechnology and biosafety to keep abreast with biotechnology developers. Information exchange systems as provided by a number of organizations fulfil this need by facilitating international cooperation, but can only be used by developing countries where the appropriate expertise exists. Even more limiting, several of these information networks are difficult to search, while others are limited in scope. The Intergovernmental Committee for the CPB, realizing the capacity constraints of developing countries, has established a coordination mechanism to maximize synergies, complementarity and collaboration between the numerous international initiatives. The evaluation of food is not only about science. It should also take into account the social, ethical and religious concerns of the local populations.

HARMONIZATION

At the international level, protocols have been agreed upon that implicitly promote the harmonization of regulatory systems. While the Codex *Principles for the risk analysis of foods derived from modern biotechnology* are available to guide the safety assessment of GM food, they have no binding effect on national legislation, but do form the basis for harmonization under the SPS Agreement. On the other hand, the CPB has established legally binding rules for environmental risk assessments. In addition, OECD has experience in promoting international harmonization in the regulation of biotechnology by ensuring efficiency in the evaluation of environmental and human-health safety, through its working group for harmonization in biotechnology and its task force for the safety of novel foods and feeds. Developing countries therefore have sets of agreed principles for guidance, and the advantage of learning from the experiences of their forerunners by investigating best practices and adapting them to suit their individual situations. Although agreement has been reached on the scientific principles of food-safety assessment, consensus has not been achieved on the extent of data required to comply with these principles or on the role of the data in decision-making. Harmonization of components of the scientific review process has a potential benefit where a lack of resources threatens effectiveness, and the affected countries in the region have determined and agreed on the regulatory objectives.

The advantages of regional/subregional cooperation are to facilitate regulation, promote the sharing of resources, synchronize the assessment of foods derived from modern biotechnology, and expedite information exchange. The Nuffield Council of Bioethics recommends the implementation of international standards and the sharing of risk-assessment methodologies and results, particularly between developing countries with similar ecological environments. Moreover, integrating some activities could reduce the overall requirement for new financial resources. Harmonization can be achieved at several levels, *i.e.* some elements of the framework can be implemented at the regional level.

The countries of the Association of South-East Asian Nations have come together to cooperate on various levels, including:

- Harmonization of legislation for products derived from modern biotechnology and intellectual property rights;
- R&D in biotechnology; and
- Environmental protection.

ASEAN is also looking at a regional approach to biosafety, although it is not clear what is intended, *i.e.* whether regional assessment and national decision-making would be considered. Those countries in the region that have made some progress have gone as far as developing labelling regulations, although they acknowledge that implementation may not be possible in the near future due to a lack of human resources. After the 2002 humanitarian crisis in southern Africa, where a number of countries experiencing severe drought and food shortages questioned the use and safety of GM food aid, a Council of Ministers of the Southern African Development Community established an Advisory Committee on Biosafety and Biotechnology to develop a common position on biotechnology and harmonize biosafety legislation in the region.

The objective is to facilitate the movement of food products that may contain GM material across the region in future. Although harmonization may absorb some of the costs that could be incurred in establishing regulatory frameworks, the flexibility allowed by international agreements creates room for divergence from the basic principles. Also, none of the regimes give guidance on regulations. Therefore, achieving harmonization in this context may be debatable as countries grapple with criteria for the precautionary approach and socioeconomics.

Nevertheless, particular attention should be paid to supporting and creating new strategic partnerships. Countries need to find effective ways of working together, and analyse the benefits and costs of harmonization.

GM FOOD AND FOOD SECURITY

WHAT IS FOOD SECURITY

The official definition of food security, adopted at the World Food Summit of 1996, states:

- "Food security exists when all people, at all times, have physical and economic access to sufficient, safe and nutritious food to meet their dietary needs and food preferences for an active and healthy life."

This definition is understood within the framework of sustainability and was drawn from chapter 14.6 of Agenda 21, adopted at the 1992 United

Nations Conference on Environment and Development, which states: "The major thrust of food security...is to bring about a significant increase in agricultural production in a sustainable way and to achieve a substantial improvement in people's entitlement to adequate food and culturally appropriate food supplies." The underlying assumption is that the means of increasing food availability in many countries exist, but are not being realised because of a range of constraints. In identifying and resolving these constraints, it is necessary to find sustainable ways of improving and reducing year-to-year variability in food production and open the way for broader food access.

The causes of food insecurity involve a complex interplay of economic, social, political and technical issues. An analysis of this interplay should determine the potential solution and best approach for a given population group. The issue for some communities is being able to produce sufficient food. For others, lack of money to purchase a wider selection of foods is the problem. Food insecurity and poverty are strongly correlated. The Swedish International Cooperation Agency defines poverty as a three-fold deficiency: a lack of security, ability and opportunity. Poverty is the main cause of food insecurity, and hunger is also a significant cause of poverty. Hunger is not only about quantity—it goes hand-in-hand with malnutrition. Food insecurity and malnutrition impair people's ability to develop skills and reduce their productivity.

A lag in farm productivity is closely associated with rural poverty and hunger. Food insecurity nevertheless is a reality experienced by the vulnerable in all societies and in all countries, developed and developing. In developed countries, the problem of food security is often a reflection of affordability and accessibility through conventional channels. Food security for the rural poor in developing countries is about producing or securing enough to feed one's household and being able to maintain that level of production year after year. Hunger and malnutrition increase susceptibility to disease and reduce people's ability to earn a livelihood. In instances where hunger is related to household income, improving food security by ensuring access to food or increasing the purchasing power of a family is essential. Providing poor communities with the skills to improve conditions in an economically and ecologically sustainable manner creates a window of opportunity to alleviate poverty at the subsistence-farming level and, on a larger scale, by having an impact on the economic development of the country.

THE CHALLENGES TO FOOD SECURITY

In developing countries, 800 million people are undernourished, of which a significant proportion live on less than US$1 per day, despite a more than 50% decline in world food prices over the past 20 years. Global food production has soared, making a variety of foods available to all consumers. Although the decline in food prices in developed countries has benefited the poor who spend a considerable share of their income on food, this trend has not had much impact on the majority in the developing world, with sub-Saharan Africa painting the gloomiest picture.

Due to the substantial price reduction in this commodity sector, cereals have become the staple foods in the diet of poor people. While yield increases in the major cereal crops has meant more calorific intake of food, micronutrient malnutrition remains a serious problem. Regional analyses depict sub-Saharan Africa as the only region where both the number and proportion of malnourished children have consistently risen in the past three decades. However, malnutrition in South Asia is also very high.

The world population is projected to reach 8 billion by 2025, and it is estimated that most of this growth will occur in developing countries. Feeding and housing an additional 2 billion people will cause considerable pressure on land, water, energy and other natural resources. Looking at projections to 2020, the worldwide per capita availability of food is expected to increase by approximately 7%, *i.e.* 2900 calories per person per day. Nonetheless, an average availability of 2300 calories is projected for individuals in sub-Saharan Africa, a figure that is just above the recommended minimum calorie intake for an active and productive life. In terms of agricultural output, preliminary world estimates for 2001 suggested that growth was as low as 0.6%. Annual rates also demonstrate a trend of decreased productivity, particularly in developing-country regions. Output growth in Asia has systematically declined over the past five years the rates experienced in sub-Saharan Africa are lower than average. Agricultural productivity is important for food security in that it has an impact on food supplies, prices, and the incomes and purchasing power of farmers. Improving food security at the national level requires an increase in the availability of food through increased agricultural production, or by increasing imports.

To augment domestic production and maintain an adequate supply of food, food-insecure countries often rely on imports and food aid. Export earnings are frequently low and do not suffice to provide foreign exchange to finance imports. Thus, in the long term, importing food is unsustainable. Historically, increased food production in the developing countries can be attributed to the cultivation of more land rather than to the deployment of improved farming practices or to the application of new technologies. By its very nature, agriculture threatens other ecosystems, a situation that can be exacerbated by over-cultivation, overgrazing, deforestation and bad irrigation practices. However, increased demands for food in Asia, Europe and North Africa have to be met by increasing yields because most land in these areas is already used for agriculture.

The potential to expand agricultural land exists in Latin America and sub-Saharan Africa only, where much of the remaining land is marginal for agricultural expansion. The implication, therefore, is that the increase in food production needed to feed the world's growing population can only be met by increasing the amount of food produced per hectare. Recognizing the extent of environmental degradation caused mainly by human activities, the multilateral agreements that arose from the UNCED meeting of 1992 were intended to address the compromised food-security situation on a global scale. One such agreement is the United Nations Convention to Combat Desertification. This agreement promotes the implementation of practices intended to reverse desertification for sustainable land use and food security. As the more affluent developed countries tend to produce more food, some argue that redistribution of these surpluses could feed the escalating populations of developing countries. Redistribution, however, requires policy changes that may be impossible to implement on a global scale. Therefore, a substantial proportion of the food demands of developing countries will have to be met by the agricultural systems in these countries. Enabling a consistent and sustainable supply of food will require an overhaul in the production processes and the supporting infrastructure. Finding solutions to declining crop yields requires an effort that will improve the assets on which agriculture relies; namely, soils, water and biodiversity.

Transforming the agricultural systems of rural farmers by introducing technologies that integrate agro-ecological processes in food production, while minimizing adverse effects to the environment, is key to sustainable agriculture. In addition, increases in crop yield must be met with the use of locally available low-cost technologies and minimum inputs without causing damage to the environment.

ATTAINING FOOD SECURITY

Within the context of the definition, three distinct components appear central to attaining food security: availability, accessibility and adequacy. Within each component, questions are raised which may need to be addressed to improve the food security situation at a national, regional or international level.

The questions raised here are intended to demonstrate the complexity of the issue and are by no means exhaustive:

- *Availability*: is there enough food available through domestic production or imports to meet the immediate needs? Is production environmentally sustainable to meet long-term demands? Are the distribution systems effective in reaching low-income and rural communities?
- *Accessibility*: do the vulnerable in society have the purchasing power to attain food security? Can they afford the minimum basic diet of 2100 calories per day required for an active and productive life?
- Adequacy: does the food supply provide for the differing nutritional needs, *i.e.* a balanced diet, offering the necessary variety of foods at all times? Is the food properly processed, stored and prepared?

Global food productivity is undergoing a process of rapid transformation as a result of technological progress in the fields of communication, information, transport and modern biotechnology. A general observation is that technologies tend to be developed in response to market pressures, and not to the needs of the poor who have no purchasing power. As agriculture is the main economic activity of rural communities, optimizing the levels of production will generate employment and income, and thus uplift the wealth and well-being of the community.

Improving agricultural production in developing countries is fundamental to reducing poverty and increasing food security. Investment to raise agricultural productivity can be achieved through the introduction of superior technologies such as better-quality seeds, crop rotation systems etc. It is argued, however, that the adoption of earlier agricultural technologies has led to the emergence of more virulent strains of pests, pathogens and weeds, soil deterioration and a loss of biodiversity.

The Green Revolution, in particular, focused on wheat and rice—not much attention was paid to staple crops such as sorghum, cassava or millet. Also, the seeds and fertilizers required to grow the higher-yielding varieties were expensive and therefore not accessible to all. Reaffirming support for the principles agreed upon at the UNCED, the United Nations Millennium Development Goals have set a road map for protecting the environment. Embraced in these time-bound goals is a new development ethic that demands sustainability in a framework where progress is measured in terms of actions reconciling the economic and ecological factors of food production for the benefit of present and future generations?

Extending this understanding to agriculture, sustainable agriculture is defined as:

- Environmentally sound, preserving resources and maintaining production potential;
- Profitable for farmers and workable on a long-term basis;
- Providing food quality and sufficiency for all people;
- Socially acceptable; and
- Socially equitable, between different countries and within each country.

A secure food system is one in which the ecological resources on which food production depends allow for their continued use, with minimum damage for present and future generations.

In other words, food security and sustainable agriculture are interlinked and both are central to the concept of sustainable development. The FAO Anti-hunger Programme reported that increasing investment in agriculture and rural development can reduce hunger. To reduce the number of hungry people by half by 2015, it estimated that funding of US$24 billion would be required for agricultural research, emergency food assistance, and improving rural infrastructure.

By contrast, at the current rate of progress, the number of food-insecure would fall by only 24%. For people to be food-secure, they must have access to the resources needed to buy or produce their own food. Breaking the poverty cycle of rural communities, whose livelihoods depend on agriculture, will require investment in different technologies to address the different constraints experienced in different world regions. The production problems experienced by farmers vary between countries and communities, and technological solutions need to be relevant to those circumstances, *i.e.* one solution will not be suitable everywhere.

The potential of some of these technologies has been demonstrated in various world regions— for instance, agro-ecological improvement programmes involving:

- Better harvesting and conservation of water, even in rain-fed environments;
- A reduction of soil erosion by adopting zero tillage combined with the use of green manure and herbicides as in Argentina and Brazil; and
- Pest and weed control without pesticide or herbicide use, *e.g.* Bangladesh and Kenya, has been well tested and established.

Indeed, such programmes are now widely accepted as being at the core of sustainable agriculture. The communities that participated in these projects were able to transform food production through the use of resource-management strategies that focused on improving the soil by growing leguminous crops and applying agro-forestry, zero tillage and green manure. These and other projects have proved that the sustainability of any farming practice and the conditions under which production can be maintained at reasonable levels cannot be predicted with absolute certainty. Some regions may be better able to transfer high-yielding technologies with varying degrees of success.

The uptake of new production systems has proved successful where the programmes have included the participation of entire communities and have not been introduced to isolated groups of farmers. Producing nutritionally enhanced properties in staple crops eaten by the poor could reduce the burden of disease in many developing countries. Scientists at the International Crops Research Institute for the Semi-Arid Tropics have developed a pearl millet variety enhanced with beta-carotene. The trait naturally occurs in two Burkina Faso millet lines from which it was transferred by conventional breeding methods. Genetic modification of japonica rice with a ferritin gene has not given superior results compared to rice with an 80% increase in iron density produced by conventional plant breeding at the International Rice Research Institute. Research and technology alone will not drive agricultural growth. Inadequate infrastructure and poorly functioning markets tend to exacerbate the problem of food insecurity. The cost of marketing farm produce can be prohibitive for small-scale farmers, as their isolation prevents the link between agricultural and non-agricultural activities among adjacent villages and between rural and urban areas. Building roads in rural areas is vital to facilitating growth, trade and exchange of farm and non-farm products in rural communities, even those that can adequately feed themselves. For instance, government investment in irrigation projects, storage and transport facilities, roads connecting villages to larger markets in the rural areas of China and India, has made an impressive impact on employment and productivity and ultimately provided opportunities for poverty alleviation in the affected areas. UNDP, basic thresholds in roads, power, ports and communications must be reached in order to sustain growth.

GENETIC TAGGING OF AQUACULTURAL SPECIES

Species-specific DNA markers can be used to identify animal species such as commercial molluscs and crustaceans, which represent a high proportion of aquacultural species. For instance, in Thailand, at the National Centre for Genetic Engineering and Biotechnology Marine Biotechnology Unit, species-specific markers based on 16S ribosomal DNA polymorphism have been developed for penaeid shrimps, tropical abalone and oysters. The black tiger shrimp is the most commercially important cultured species in Thailand. Because of outbreaks of diseases, the white shrimp has been introduced into Thailand and cultured commercially. On the other hand, external characteristics of P. monodon and P. semisulcatus are similar, but the growth rate of the latter is approximately three times slower than that of the former. In addition, P. merguiensis larvae, which could not yet be successfully cultured, were sold as those of P.vannamei. Species-specific markers were therefore developed for identifying the afore-mentioned species and P. japonicus as well. These markers can be applied to ensure quality control by properly labeling traded shrimp larvae.

Three species of tropical abalone are found in Thailand's waters: Haliotis asinina, H. ovina and H. varia. However, H. asinina is the most productive one, as it provides the highest ratio between meat weight and total weight. H. asinina specific markers based on 16S rDNA polymorphism have been developed in order to prevent supplying the wrong abalone larvae for the industry as well as to foster quality control of abalone products from Thailand.

Oyster farming has shown rapid growth in Thailand over the last few years. Taxonomic difficulties relating to Thai oysters have had a limiting effect on the culture efficiency and development of their closed life cycle. Molecular genetic markers have therefore been developed to identify the three commercially cultured oysters, Crassostrea belcheri, C. iredalei and Saccostrea cucullata.

DNA FINGERPRINTING OF GRAPEVINE VARIETIES

Since 1990, US and French researchers have been trying to establish the phylogenetic tree of the varieties of grapevine grown throughout the world, using the fingerprinting technique. The research consists of analyzing the structure of certain regions of the genome of some grapevine varieties and comparing it with that of other varieties, so as to establish possible phylogenetic relations. For this analysis, the DNA is extracted from young ground leaves, but also from fruits and branches. The results of this first research have been published in 1999 in the Science journal.

There are about 6,000 grapevine varieties cultivated worldwide. In order to establish the origin of a variety, specialists used to rely on phenotypic traits, such as the morphology of leaves, berries and grapes. In this way, varieties could be grouped in a few families. DNA fingerprinting enables the researchers to go further. It appears that the current grapevine varieties are the remote offspring of the grapevines grown during the Antiquity around the Mediterranean, or during the European Middle Ages. The current varieties are the result of lengthy breeding activities, identification and comparison work, and stabilization of the selected strains or lines. The team led by C. Meredith and J. Bowers of the Department of Viticulture and Oenology of the University of California, Davis, has confirmed that the cabernet sauvignon variety – which is dominant in the Medoc region of France and is at the origin of most red grapevine varieties in the New World – was in fact the offspring of the cabernet franc and sauvignon, two varieties deeply entrenched in the middle valley of the Loire river.

The cooperation between the Californian team and the French specialists of the National Higher School of Agronomy, Montpellier, associated with the Genetic Research and Breeding Unit of the National Agricultural Research Institute has led to undisputable results concerning the origins of various grapevine varieties: the chardonnay, aligote, gamay and melon of Burgundy are all cousins and almost siblings. Such a conclusion, derived from the analysis of DNA fingerprinting, was not a surprise for oenologists and tasters because the wines made from these four varieties share common structures and aromas. This kinship is particularly expressed in the ageing wines. Similarly, aged wines of the chenin variety resemble the Hungarian tokay. But more surprising than the discovery of that kinship among the four grapevine varieties, was the identity of one of the progenitors of the initial couple that gave rise to these varieties. Indeed, several historical elements were in favour of the creation of lines through the cross pollination between the pinot noir and the white gouais; these crosses have given birth – as proven by the fingerprinting analysis – to the three white and red varieties grown for a long time in various French provinces, and for some decades, in many regions of the globe.

The surprising aspect of this discovery was that the white gouais is almost unknown, although the vine specialists in Montpellier continue to grow it and to make wine form it for their own pleasure. However, this variety has played a key role in the origin of French viticulture, just as to R. Dion in his Histoire de la vigne et du vin en France des origines au XIXe siècle. A document dated from the 12th century mentioned this variety as a lower-grade one; in 1338, the white gouais was found in Metz under the name of goez; at that time, instructions were given to eliminate this variety from all the Metz territory and to privilege only the white and black fromental, considered as higher-grade varieties.

The gouais was found in Paris during the 14th century and, owing to the expansion of the workers' population, it progressively replaced the pinot noir of Burgundy, which was a good variety of Parisian vineyards. The extension of the gouais was due to the wish of the winemakers to produce a cheaper wine. However, the phenomenon was limited to Paris and its suburbs; in the vineyards located away from the capital, the gouais was rejected, more noble grapevine varieties were used and contributed to the reputation of French viticulture.

The white gouais was also formerly grown in the Jura and Franche-Comté. For the US and French researchers, this variety which has played a key role in the history of vine and wine, is the same as the heunisch variety of Central Europe, introduced in Gaul by a Roman emperor originating from Dalmatia. In Montpellier, the French researchers participating in the joint study with the US scientists from the University of California, Davis, have tried to reproduce the breeding between the pinot noir and white gouais in order to seek confirmation of the genetic research. Other attempts were expected to widen even more the range of cultivated grapevine varieties, for both their fruits and wines derived from them. But this approach was hindered by a drastic regulation, which practically prohibits any venture of this kind, while non-French winemakers and vinegrowers could do it.

In Apulia, in the heel of the Italian boot, drawing on grapes grown by up to 1,600 small farmers in the area, a California wine consultant associated with another Italian wine consultant from Friuli are producing and marketing wines that have scored a great success worldwide, with 2004 projected sales 15 times as big as those in 1998, the winery's first year. The wines are called A-Mano – handmade – and by far the best known is a robust red made from a once-obscure grape named primitivo. DNA testing by Carole Meredith at the University of California, Davis, established that primitivo is a descendant of a grape called crljenak kastelanski, widely known in the 18th and 19th centuries on the Dalmatian coast of Croatia.

California's zinfandel, she showed, is genetically the same as primitivo, though how it crossed the ocean remains a subject of dispute. Apulian primitivo and zin are not twins, of course; climate, soil and vinification all help to shape a wine's look, aroma and flavour, along with the grape variety. But the two share several characteristics: both are fruit-rich, chewy, sometimes lush wines, a deep violet-red in colour, often too high in alcoholic content for comfort, but much more subtle if carefully handled.

For years, primitivo was used to add unacknowledged heft to chianti, barbaresco and even red burgundy. Nowadays, primitivo can stand on its own feet. In addition to A-Mano primitivo, other high-quality primitivos are grouped in an organization called the Academia dei Racemi, not a true cooperative but an association in which each member makes his own wine and joins the others for marketing support and technical advice. Based in Manduria, between the old cities of Taranto and Lecce, the group includes value-for-money labels like Masseria Pepe, Pervini and Felline.

BIOFORTIFICATION OF FOOD CROPS

Biofortification of food crops makes sense as part of an integrated food-systems approach to reducing malnutrition. It addresses the root causes of micronutrient deficiencies, targets the poorest people, and is scientifically feasible and cost-effective. It is a first step in enabling rural households to improve family nutrition and health in a sustainable way. HarvestPlus is a coalition of CGIAR Future Harvest Centres or Institutes, partner collaborating institutions and supportive donors. The International Centre for Tropical Agriculture and IFPRI are coordinating the plant breeding, human nutrition, crop dissemination, policy analysis and impact activities to be carried out at international Future Harvest Centres, national agricultural research and extension institutions, and departments of plant science and human nutrition at universities in both developing and developed countries.

An initiative of the Consultative Group on International Agricultural Research, HarvestPlus is a global alliance of research institutions and implementing agencies coming together to breed and disseminate crops with improved nutritive value, *e.g.* with a higher content of iron, zinc and vitamin A. The biofortification approach is backed by sound science.

Research on this funded by the Danish International Development Assistance and coordinated by the International Food Policy Research Institute led to the following conclusions:

- Substantial, useful genetic variation exists in key staple crops;
- Breeding programmes can readily manage nutritional quality traits, which for some crops have proven to be highly suitable and simple to screen for;
- Desired traits are sufficiently stable across a wide range of growing environments; and
- Traits for high nutrition content can be combined with superior agronomic traits and high yields.

Initial biofortification efforts will focus on six staple crops for which prebreeding studies have been completed: beans, cassava, sweet potatoes, rice, maize and wheat. The potential for nutrient enhancement will also be studied in ten additional crops that are important components in the diets of those with micronutrient deficiencies: bananas/plantains, barley, cowpeas, groundnuts, lentils, millet, pigeon-peas, potatoes, sorghum and yams. During the first four years of the project, the objectives are to: determine nutritionally optimal breeding objectives; screen CGIAR germplasm for high iron, zinc and beta-carotene amounts; initiate crosses of high-yielding adapted germplasm for selected crops; document cultural and food-processing practices, and determine their impact on micronutrient content and bioavailability; identify the genetic markers available to facilitate the transfer of traits through conventional and novel breeding strategies; carry out in-vitro and animal studies to determine the bioavailability of the enhanced micronutrients in promising lines; and initiate bio-efficacy studies to determine the effect on biofortified crops on the micronutrient status of humans.

During the following three years, the objectives are to: continue bio-efficacy studies; initiate farmer-participatory breeding; adapt high-yielding, conventionally-bred, micronutrient-dense lines to select regions; release new conventionally-bred biofortified varieties to farmers; identify gene systems with potential for increasing nutritional value beyond conventional breeding methods; produce transgenic lines at experimental level and screen for micronutrients, test for compliance with biosafety regulations; develop and implement a marketing strategy to promote the improved varieties; and begin production and distribution. During the last three years of the project, production and distribution of the improved varieties will be scaled up; the nutritional effectiveness of the programme will be determined; and the factors affecting the adoption of biofortified crops, the health effects on individuals and the impact on household resources will be identified.

Rice

Rice is the dominant cereal crop in many developing countries and is the staple food for more than half of the world's population. In several Asian countries, rice provides 50 per cent to 80 per cent of the calorie intake of the poor. In South and South-East Asian countries, more than half of all women and children are anaemic; increasing rice nutritive value can therefore have significant positive health impact. Food-consumption studies suggested that doubling the iron content in rice could increase the iron intake of the poor by 50 per cent; germplasm screening indicated that a doubling of iron and zinc content in unmilled rice was feasible. Milling losses vary widely by rice variety, with losses of iron being higher than losses of zinc, which suggests than more zinc is deposited in the inner parts of the rice endosperm. Under the HarvestPlus project, improved rice germplasm will be provided to national partners in Bangladesh, Indonesia, Vietnam, India and the Philippines. The improved features will be incorporated into well-adapted and agronomically-preferred germplasm in ongoing breeding programmes at the national and regional level. A plant-biotechnology approach is the current priority for enhancing provitamin-A content of the rice endosperm. The leading varieties will be field tested for agronomic performance and compositional stability in at least four countries.

Wheat

The International Maize and Wheat Improvement Centre is leading the HarvestPlus research endeavour on wheat biofortification in order to increase people's intake of iron and zinc. Given that spring wheat varieties developed by CIMMYT and its partners are used in 80 per cent of the global spring wheat area, the potential impact of iron-enhanced wheat could be dramatic. The initial target countries will be Pakistan and India, in the area around the Indo-Gangetic plains, a region with high population densities and high micronutrient malnutrition.

The highest contents of iron and zinc in wheat grains are found in landraces of wild relatives of wheat such as Triticum dicoccon and Aegilops tauschii. Because these wild relatives of wheat cannot be crossed directly with modern wheat, researchers facilitated the cross between a high-micronutrient wild relative, Aegilops tauschii, and a high-micronutrient primitive wheat, Triticum dicoccon, to develop a variety of hexaploid wheat that can be crossed directly with current modern varieties of wheat and have 40 per cent to 50 per cent higher contents of iron and zinc in the grain than modern wheat.

The first biofortified lines will be delivered to the target region by 2005, *i.e.* broadly-adapted, high-yielding, disease-resistant wheat lines. The first high-yielding lines with confirmed iron and zinc contents in the grain should be available for regional deployment by mid-2007. Researchers will be exploring the introduction of the ferritin gene in wheat and will establish the feasibility of increasing the concentration of iron and zinc in the grain using advanced biotechnology approaches in addition to conventional plant breeding. Molecular markers for the iron and zinc genes that control concentration in the grain were being identified in order to facilitate their transfer. Scientists will also carry out studies on bioavailability to determine the extent to which iron and zinc status in animal and human subjects is improved when biofortified varieties are consumed on a daily basis over several months.

Maize

Maize is the preferred staple food of more than 1.2 billion consumers in sub-Saharan Africa and Latin America. Over 50 million people in these regions were vitamin A-deficient in 2004. The International Maize and Wheat Improvement Centre and the International Institute of Tropical Agriculture are identifying micronutrient-rich maize varieties and will carry out adaptive breeding for local conditions in partnership with National Agricultural Research Systems in Africa and Latin America.

The project under HarvestPlus is initially focusing on maize varieties having increased contents of provitamin A because a useful range of genetic variation has already been identified for this trait. The first target countries are Brazil, Guatemala, Ethiopia, Ghana and Zambia. To support the breeding programme, research is being conducted in Brazil, the USA and Europe to develop simple, inexpensive and rapid screening protocols for provitamin A, so as to reduce the cost of assays from $70-100 to $5-10 per sample. Research in Brazil and the USA is also focused on finding genetic markers to facilitate marker-assisted selection for provitamin A concentration. In collaboration with the University of Wageningen, a human efficacy trial was planned with provitamin A-rich maize in Nigeria for 2005 in order to study provitamin A retention or loss for different storage, processing and common cooking methods. To facilitate extension and dissemination of biofortified maize varieties, country teams will be formed in the target countries in order to conduct adaptive breeding research, farmer-participatory variety evaluations, nutritional advocacy and promotional activities.

Beans

Common beans are the world's most important food legume, far more so than chickpeas, faba beans, lentils and cowpeas. For more than 300 million people, an inexpensive bowl of beans is the main meal of their daily diet. The focus of HarvestPlus research is on increasing the concentration of iron and zinc in agronomically superior varieties. Over 2,000 accessions from the International Centre for Tropical Agriculture gene-bank and several hundred collections of African landraces have been screened for their nutrient contents. While the average iron concentration in these varieties is about 55 mg per kg, researchers have found varieties the content of which exceeds 100 mg per kg. The eventual goals are to obtain favourable combinations for productivity and nutritional traits, double the iron concentration and increase zinc concentration by about 40 per cent. The first bred lines with 70 per cent higher iron will likely emerge in 2006, while lines with double concentration of iron are anticipated in 2008.

The proportion of iron and zinc that can be absorbed from legumes such as common beans is typically low, due to anti-nutrients, specifically phytates and polyphenols, which normally bind to the iron and zinc, making them unavailable to the organism. Research indicated that it might be possible to reduce polyphenol concentrations genetically, thereby improving iron bioavailability. In contrast, vitamin C is an iron-absorption enhancer because it binds to iron and prevents it from becoming attached to the iron-absorption inhibitors. Beans are often consumed with vegetables, including bean leaves with the potential of bean leaves as a source of vitamin C still to be explored.

Cassava

Cassava, also known as manioc or tapioca, is a perennial crop native of tropical America that is also widely consumed in sub-Saharan Africa and parts of Asia. With its productivity on marginal soils, ability to withstand disease, drought and pests, flexible harvest dates, cassava is a remarkably adapted crop consumed by people in areas where drought, poverty and malnutrition are often prevalent. Cassava is typically white in colour and, depending on the amounts of cyanogenic compounds, can be sweet or bitter. The International Centre for Tropical Agriculture will coordinate HarvestPlus' overall activities on cassava biofortification and be primarily responsible for research in Asia, Latin America and the Caribbean.

The IITA will be responsible for cassava biofortification in Africa. In collaboration with the University of Campinas, São Paulo State, Brazil, the total content of provitamin A in cassava varieties will be determined spectrophotometrically. Provitamin-A retention studies will also be carried out on different preparation and cooking methods used in cassava-consuming countries.

A method for storing cassava roots for several weeks or a few months is needed for programmes quantifying hundreds of samples per year. Initial data suggest that the anti-oxidant property of a few yellow pigments in cassava roots may delay physical deterioration of the roots.

The longer shelf life of yellow cassava roots may not only appeal to farmers and consumers, but may also increase the demand for biofortified varieties. Nutritionally improved germplasm coupled with superior agronomic performance can be developed as a medium-term approach with products reaching the farmers as soon as 2009. The aim is to identify and select, from the varieties having both high provitamin-A contents and good agronomic performance, those with the highest iron and/or zinc content.

Sweet Potato

Sweet potato is an important part of the diet in East and Central Africa where vitamin-A deficiency is widespread. At present, African predominant sweet potato cultivars are white or yellow-fleshed varieties that contain small amounts of provitamin A. In contrast, the orange-fleshed varieties are believed to be one of the least expensive, rich, year-round sources of provitamin A. Boiled orange-fleshed sweet potato, such as the Resisto variety developed in South Africa, contains between 1,170 and 1,620 Retinol Activity Equivalents per 100 g and is estimated to provide between 25 per cent and 35 per cent of the recommended daily allowance for a preschool child. Experts at the International Potato Centre who developed a biofortified orange-fleshed sweet potato, estimated that when fully disseminated, this sweet potato could reduce vitamin-A deficiency in as many as 50 million children. To encourage a switch from non-orange to orange-fleshed varieties, the texture of the latter must be changed because they tend to have a high-moisture content and adults prefer varieties with a low water content, *i.e.* a high dry biomass. Plant breeding is ongoing to increase the dry biomass of the provitamin A-rich orange varieties, to improve organoleptic characteristics and at the same time improve their resistance to viruses and drought. About 40 varieties of sweet potato with high dry biomass and provitamin-A content have been introduced to sub-Saharan Africa. Of these, 10 to 15 were being tested widely in different agro-ecological areas in some countries.

Some original varieties, mainly local landraces, have been well accepted by farmers and were being distributed on a small scale. HarvestPlus' biofortification activities in sweet potato will be initially focused on Ethiopia, Ghana, Kenya, Mozambique, Rwanda, South Africa, Tanzania and Uganda. The variation in provitamin-A content of newly harvested roots can be as much as 45 per cent. Much of the provitamin A appears to be retained during storage, food preparation and cooking. In the South African Resisto variety, the provitamin-A activity of the boiled roots was between 70 per cent and 80 per cent of that of freshly harvested roots. Additional studies were to be carried out in 2004 to determine the provitamin-A losses during food processing and cooking based on the usual practices found in East and Central Africa. A human bioefficacy study using an organoleptically acceptable promising variety was planned for 2005, once the food processing studies were completed.

The $100-million ten-year HarvestPlus programme will be financed during the first four years mainly by the World Bank, the USAID and DANIDA. The Bill and Melinda Gates Foundation would contribute $25 million towards the total cost of the programme. In addition, the Canadian Agency for International Development will allocate funds for the Latin American part of the programme.

INDUSTRIAL PRODUCTION OF HEALTHIER FOODSTUFFS

Food science and biotechnology can lead to substantial innovations in the production of healthier foodstuffs as well as increased profits by major food companies as in the period 2003-2004. Consider Nestlé. The group is selling beverages, mineral water, dairy products, ice-creams, precooked meals, chocolate, pet food and cosmetic products. In 2002, Nestlé's annual turnover amounted to 87.7 billion Swiss Francs, broken down as follows: beverages, including mineral water; dairy products; precooked meals; confectionery; pet food; cosmetic products.

However, in 2003, net profit decreased to 6.2 billion SF, 17.3 per cent less than in 2002, owing to the weak economic growth in Europe and monetary fluctuations. Nestlé's biggest market is Europe with sales of 28.5 billion SF, followed by the American markets. The Asia-Pacific region also became a priority for the group's development with the turnover in that region reaching 14.4 billion SF in 2003.

Nestlé spends 1.35 per cent of sales on research and development – a lot for a food company – and was employing 250,000 persons in 2003-2004 worldwide. It explores the frontiers of nutrition research to determine what people should and should not be eating, to develop products such as milk with added long-chain polyunsaturated fatty acids and non-dairy products fortified with calcium for the lactose-intolerant individuals. Yakult – a bland, sweet, yellowish drink – is also a good example of industry that made good in healthy drinks. It is produced by the Japanese company Shirota, founded in 1955. Minoru Shirota discovered Lactobacillus casei shirota in 1930.

The product was launched in Europe in 1994 and since then has spread across the world. It claims the beneficial effects of lactobacilli on the intestinal microflora. It represented a $2 billion global business, and encouraged competition from other companies. Cargill, Inc., whose core business is commodities, employed 200 food scientists in 2003, up from 20 in 2000. It has developed many products with new ingredients, including Bon Appétit, a raspberry tea with soybean isoflavones, which 'may help promote bone health and relieve some of the symptoms of menopause'. While Kraft Foods and Cadbury Schweppes claimed they were removing some of the trans-fats out of their foodstuffs, PepsiCo, Inc., stated it has taken all the trans-fats out of its Frito-Lay snacks. This move was to a large extent the cause of a 30 per cent boost in fourth-quarter earnings.

The drinks and snacks maker's quarterly profit was also lifted by lower costs associated with its 2001 merger with Quaker Oats. Fourth-quarter earnings were $897 million, or 51 cents a share, compared with $689 million, or 39 cents a share, in the same quarter a year earlier. Revenue rose 9.4 per cent to $8.1 billion. The company continues to expand its snacks line with healthier offerings, *e.g.* new crisps, using maize oil rather than oil containing trans-fats. Frito-Lay's North American sales grew 6 per cent to $2.7 billion in the fourth quarter with volume up a smaller 3 per cent. The unit controlled almost two-thirds of the US snacks market. PepsiCo, Inc., is the world's fourth-biggest agri-food group, behind Nestlé, Kraft Foods and Unilever. In 2003, its turnover reached $26.971 billion and its net profit was $4.781 billion. Present in 160 countries, it had 140,000 employees.

The modification of vegetable oils is one of the key areas of plant and crop biotechnology, the overall objective being to increase their content in unsaturated fatty acids and to decrease that of saturated ones through conventional breeding, induced mutations or genetic engineering. Extensive work has been carried out on oilseed rape, soybeans, peanut and sunflower with good results that led to the commercialization of several products. Palm oil, which contains an equal proportion of saturated and unsaturated fatty acids, in addition to beta-carotene, is also a current research target, particularly of researchers at the Palm Oil Research Institute of Malaysia. In addition, replacing triglycerides with diglycerides in vegetable oils render them free of trans-hydrogenated fats and good cooking oils, *e.g.* 'econa oil' in Japan. Inulin and oligofructosans refer to a group of fructose-containing carbohydrate polymers which, in many plant species, act as protective agents against dehydration and cold temperatures and also offer many health benefits to humans, mainly in the stimulation of the growth of beneficial micro-organisms called bifidobacteria.

These bacteria are sometimes used as a probiotic additive to foodstuffs such as yoghurt, as they can defeat harmful bacteria in the intestines and produce compounds with good health benefits. These dietary fructans are also reported to have a lipid-lowering potential. They are not digested in the upper gastro-intestinal tract and therefore have a reduced caloric value. They share the properties of dietary fibres without causing a rise in serum glucose or stimulating insulin secretion. Inulin and oligofructosans can be used to fortify foods with fibre or improve the texture of low-fat foods without resulting in adverse organoleptic effects.

Most of these two products currently on the market are either chemically synthesized or extracted from plant sources such as chicory roots. Oligofructosans are shorter chain polymers, highly soluble and provide 30 per cent to 50 per cent of the sweetness of sugar, and also have the other functional qualities of sugars. In formulation, inulin forms a smooth creamy texture, which makes this compound suitable as a fat substitute. We can also cite the work of F. Georges of the Plant Biotechnology Institute. He was working on the production of inulin and oligofructosans in separate transgenic plant experiments to compare the efficiency of their fibre production. Oilseed rape which is a poor producer of inulin and oligofructosans, was used as model system. In particular, the production of two enzymes was to be evaluated: sucrose-1-fructose-1-transferase which adds a fructose moiety to a sucrose molecule, and fructan: fructan fructosyl transferase which continues to elongate the polymer by adding more fructose moieties to the chain.

The study showed that both enzymes could be used in conjunction to produce inulins and oligofructosans. Growers of nutraceutical plants need varieties with good agronomic potential and those that are consistent with the varieties in terms of germination time, height and maturity. Growers will need to be able to guarantee the quality of their natural health-beneficial products. Breeding methods can therefore be used to achieve uniform quality for clinical testing and for product development, as well as to remove these potentially harmful or otherwise undesirable compounds that are produced in the plants along with their therapeutic ones.

To meet these goals, Alison Ferrie of the Plant Biotechnology Institute was using the doubled haploid technology or "haploidy", which facilitates the development of true-breeding lines. Immature pollen grains, called microspores, were cultured to produce haploid lines, whose genetic stock was thereafter doubled. True-breeding plants were thus produced in one generation, and doubled haploid techniques reduced the time required to develop a new variety by about three to four years. At the NRC-PBI, doubled haploid technology has been developed for oilseed rape and wheat. It is being applied to a wide range of nutraceutical and herbal species. Over 80 species have been screened for embryogenic response; anise, fennel, dill, caraway, angelica and lovage have shown good potential. Haploidy could also be combined with mutagenesis to enhance the desirable components or decrease the undesirable characteristics. Mutagenizing single cells had definite advantages over seed mutagenesis. The new market for healthier foodstuffs attracts both the agri-food giants and pharmaceutical groups, so that the competition is harsh among them and the frontiers are less marked between both kinds of corporations. The competitive advantage of the food industry in this race is that it has a good knowledge of consumers' behaviour, massive marketing strategies while knowing that nutraceutics should remain tasteful and palatable if these were to be patronized by consumers.

In France, a success story was that of Danone's Actimel, launched in 1995 in Belgium in the form of a small bottle corresponding to an individual dose and commercialized in 15 countries. More than 600 million bottles had been sold worldwide in 1999, including about 100 million in France, where 9 per cent of the households of all socio-professional categories bought Actimel – dubbed the 'morning health gesture'. Others include that of the case of Eridania-Béghin Say in France in 1999, relating to food additives having an impact on cardio-vascular diseases, colon cancer, osteoporosis, diabetes, etc. which sold commercialized powder sugar enriched with 'biofibres', which boosts intestinal microflora and helps the body to naturally resist illness.

Back in Nestlé, they are also carrying out the relevant research-and-development work with the support of its 600-scientist strong nutrition centre, located in Lausanne while in May 1999, in the USA, Australia, and in Switzerland, Unilever with an international nutrition research centre at Vlaardingen, Netherlands, commercialized a 'hypocholesterol' margarine, which could help prevent the accumulation of 'bad' cholesterol. It also aimed to target markets in Europe and Brazil. In the USA, most agri-food companies have developed soups, beverages and cereals, which can help digestion and prevent cardio-vascular diseases and hypertension.

The US Food and Drug Administration has opened the way to nutraceutics, having labels carrying a health recommendation. On 21 October 1999, the FDA granted to soybeans the clearance to carry the claim 'may reduce cardiovascular risks' on their labels. This request was made by E.I. Dupont de Nemours and Co., Inc., the world's first-biggest producer of soybean products. Soya sauce and soybean paste are major foodstuffs across Asia. Industrial soybeans undergo a solid-state fermentation process using compliant stainless steel tanks instead of in conventional bamboo trays. They are also inoculated with Aspergillus oryzae selected strains that have been developed in Thailand to produce koji in higher yields and of better quality.

This technique, developed by a fermentation consortium associating the National Centre for Genetic Engineering and Biotechnology and the Department of Chemical Engineering of Kasetsart University has been successfully applied by the company Chain Co. Ltd., Bangkok, and thereafter adopted by some soya-sauce manufacturers in Thailand. The same company has succeeded in selecting the appropriate strain of Lactobacillus to replace the addition of acetic acid in order to enhance the sour taste of soya sauce.

The company produces the top quality commercial soya sauce in Thailand – the so-called First Formulation.

The Case of Long-Chain Polyunsaturated Fatty Acids

Long-chain polyunsaturated fatty acids are a research focus for nutritionists and food biotechnologists. Their beneficial effect on the functioning of the cardio-vascular system has been initially mentioned since the 1970s in the medical literature. In France, David Servan-Schreiber – a psychiatrist advocating a 'medicine of emotions' – Guérir le stress, l'anxiété et la dépression sans médicaments ni psychanalyse has stressed the role of these fatty acids as anti-depression substances. Incidentally, the author is also a shareholder of a company that sells pills containing these fatty acids these long-chain polyunsaturated fatty acids belong to two main categories: omega-3 and omega-6. Among omega-3 fatty acids, there are the alpha-linolenic acid with 18 carbon atoms, eicosapentaenoic acid with 20 carbon atoms and docosahexaenoic acid with 22 carbon atoms.

The human body cannot synthesize the ALA as well as the linoleic acid which is an omega-6 fatty acid. Omega-3 fatty acids are found in rapeseed and soybean oils, marine animals and human milk. Food-consumption surveys carried out in France have shown that the consumption of omega-3 fatty acids was insufficient and the ratio of omega 6 to omega 3 was not balanced. Although research is being carried out on the precise role of these fatty acids on human health, it is not easy for the public to have a clear view of established scientific facts and amid controversial statements. Let us look now at what maybe causing confusion among the public as regard the issue of omega-3 fatty acids. It may have begun with the study that revealed lower morbidity and mortality due to cardio-vascular of Greenland's Inuits who consume a lot of fatty fish. In France, the French Agency for Food Sanitary Safety convened a meeting of experts on the effects of omega-3 fatty acids on the cardio-vascular system. They concluded that the supplementation of daily diet with these fatty acids could have a beneficial impact on the functioning of the cardio-vascular system, as a secondary prevention measure. Morbidity and mortality reduction was indeed significant among the persons who suffered form cardio-vascular or metabolic diseases.

However, omega-3 fatty acids did not act on cholesterol; they may act on triglycerides and cell membranes, as well as on blood clotting and heart excitability; they may also have, through prostaglandins a positive effect on hypertension. The experts convened by the French AFSSA also warned against the role of the consumption of excessive quantities of omega-3 fatty acids, as they would increase cell susceptibility to free radicals. They recommended a maximum daily intake of EPA and DHA of 2g per day. Then there are also the claims on the prohibitive effects of omega-3 fatty acids on tumors. To this, the AFFSA experts concluded that all the studies carried out up to 2004 on food habits did not substantiate in humans any evidence indicating that an enrichment of the diet with precursors of omega-3 fatty acids would protect against cancer. However, research work carried out on rats has shown that a diet enriched with omega-3 fatty acids caused a 60 per cent decrease in size of mammary tumours, twelve days after radiotherapy, compared with a 31 per cent decrease in animals fed with a non-enriched diet.

Trials are expected to be carried out on humans. Given the insufficiency of evidence, the benefits of taking Omega-3 pills remain inconclusive. In view of this, the general advice is to consume fish at least twice a week. The same goes for rapeseed oil. This is sufficient to meet the daily needs of omega-3 fatty acids. It is also recommended to feed poultry with rapeseed meal rather than with sunflower meal, because the former is richer in omega-6 fatty acids. Thus, consuming this kind of poultry meat would provide enough omega-6 fatty acids.

A POTENTIAL ROLE FOR MODERN BIOTECHNOLOGY

The Convention on Biological Diversity dictates the use and application of relevant technologies as a means of achieving the objectives of conservation and sustainable use with specific reference to biotechnology. Modern biotechnology is purported, from a technical perspective, to have a number of products for addressing certain food-security problems of developing countries. It offers the possibility of an agricultural system that is more reliant on biological processes rather than chemical applications.

The potential uses of modern biotechnology in agriculture include: increasing yields while reducing inputs of fertilizers, herbicides and insecticides; conferring drought or salt tolerance on crop plants; increasing shelf-life; reducing postharvest losses; increasing the nutrient content of produce; and delivering vaccines. The availability of such products could not only have an important role in reducing hunger and increasing food security, but also have the potential to address some of the health problems of the developing world. Achieving the improvements in crop yields expected in developing countries can help to alleviate poverty: directly by increasing the household incomes of small farmers who adopt these technologies; and indirectly, through spill-overs, as evidenced in the price slumps of herbicides and insecticides. Indirect benefits as a whole tend to have an impact on both technology adopters and non-adopters, the rural and urban poor. Indeed, some developing countries have identified priority areas such as tolerances to alkaline earth metals, drought and soil salinity, disease resistance, crop yields and nutritionally enhanced crops.

The adoption of technologies designed to prolong shelf-life could be valuable in helping to reduce postharvest losses in regionally important crops. Prime candidates in terms of crops of choice for development are the so-called 'orphan crops', such as cassava, sweet potato, millet, sorghum and yam. Multinationals have found no incentive to develop these crops and have instead invested in marketable crops with high profit returns. This strategy is intended to target wealthier farmers in temperate-zone countries with the financial capacity and tradition of supporting new seed products. However, here is a potential for multinational companies to develop crops grown largely in developing countries. The investment costs are low and the potential markets considerably large. While some public-sector research institutes in developing countries are forging ahead with the application of modern biotechnology, a small number are supported by government policy and therefore follow a defined agenda.

Still other governments believe that the risks associated with modern biotechnology outweigh the benefits. Currently, the many promises of modern biotechnology that could have an impact on food security have not been realised in most developing countries. In fact, the uptake of modern biotechnology has been remarkably low owing to the number of factors that underpin food security issues. In part, this could be because the first generation of commercially available crops using modern biotechnology were modified with single genes to impart agronomic properties with traits for pest and weed control, and not complex characteristics that would modify the growth of crops in harsh conditions. Secondly, the technologies are developed by companies in industrialized countries with little or no direct investment in, and which derive little economic benefit from, developing countries. Thirdly, many developing countries do not have the necessary biosafety frameworks to regulate the products of modern biotechnology.

For example, it took over two years for the Kenyan authorities to approve the field-testing of a virus-resistant sweet-potato variety because the scientific capacity for evaluating the product was unavailable. It should be noted, however, that such delays in the approval process have also been seen in developed countries, especially during the initiation of national regulatory evaluation. However, this trend is quickly changing as a number of developing countries either adopt or develop appropriate biotechnologies or regulatory infrastructures.

A report by the International Service for National Agricultural Research states that more than 40 crops are the focus of public-sector research programmes of 15 developing countries involving disease-resistant traits in rice, potato, maize, soybean, tomato, banana, papaya, sugarcane, alfalfa and plantain. For example, the Brazilian Agricultural Research Corporation has concentrated research into genetically modified crops on disease resistance in beans, papaya and potatoes.

The research programme at the University of Cape Town focuses on the development of crops resistant to viruses and to desiccation. The university has recently had a breakthrough with maize streak-virus resistance. In

Thailand, the National Centre for Genetic Engineering and Biotechnology has supported research into disease resistance of rice, pepper and yard-long beans. therefore follow a defined agenda. Still other governments believe that the risks associated with modern biotechnology outweigh the benefits. Currently, the many promises of modern biotechnology that could have an impact on food security have not been realised in most developing countries. In fact, the uptake of modern biotechnology has been remarkably low owing to the number of factors that underpin food security issues.

In part, this could be because the first generation of commercially available crops using modern biotechnology were modified with single genes to impart agronomic properties with traits for pest and weed control, and not complex characteristics that would modify the growth of crops in harsh conditions. Secondly, the technologies are developed by companies in industrialized countries with little or no direct investment in, and which derive little economic benefit from, developing countries. Thirdly, many developing countries do not have the necessary biosafety frameworks to regulate the products of modern biotechnology.

For example, it took over two years for the Kenyan authorities to approve the field-testing of a virus-resistant sweet-potato variety because the scientific capacity for evaluating the product was unavailable. It should be noted, however, that such delays in the approval process have also been seen in developed countries, especially during the initiation of national regulatory evaluation. However, this trend is quickly changing as a number of developing countries either adopt or develop appropriate biotechnologies or regulatory infrastructures.

A report by the International Service for National Agricultural Research states that more than 40 crops are the focus of public-sector research programmes of 15 developing countries involving disease-resistant traits in rice, potato, maize, soybean, tomato, banana, papaya, sugarcane, alfalfa and plantain. For example, the Brazilian Agricultural Research Corporation has concentrated research into genetically modified crops on disease resistance in beans, papaya and potatoes. The research programme at the University of Cape Town focuses on the development of crops resistant to viruses and to desiccation. The university has recently had a breakthrough with maize streak-virus resistance. In Thailand, the National Centre for Genetic Engineering and Biotechnology has supported research into disease resistance of rice, pepper and yard-long beans.

Developing countries with limited financial and human resources need to find the right balance for investing in conventional and modern biotechnology research programmes. While alliances with the private sector may contribute to the search for new technologies, the public sector needs to focus on crops and traits in which the former may be unwilling or unable to invest. The extent to which priority is given to modern biotechnology over other research methods should be linked to a country's agricultural priorities and objectives as well as to its environmental concerns. Ultimately, investment in interventions that support good governance, the development of rural infrastructure and market access is required before any of the promises of modern biotechnology can be realised. In general, policies that stimulate economic growth and target poverty reduction may have significant bearing on the health and well-being of the population.

RESEARCH OWNERSHIP

Research is a critical part of any effort aimed at improving food production and reducing poverty. Globally, much of agricultural R&D is carried out in the public sector, thereby serving the interests of developing countries. Public research in developed countries and Latin America is mostly conducted by government institutions and universities, whereas almost all agricultural research in Africa is carried out in public institutions, including R&D, technology transfer and dissemination of improved plant varieties. In general, international agricultural research institutions form a second level of research development and technology providers in developing countries. Public research institutes have, in the past, researched and improved orphan crops, mainly for donation to poor farmers or at cost. Generally, academic institutions are perceived as producers of knowledge that benefit and protect the public. Also, national and international research institutes aim to address the agricultural problems of resource-poor farmers in developing countries, *e.g.* increasing productivity through the use of a variety of techniques, including modern biotechnology. In reality, public institutions are now exposed to the forces of globalization and compelled to compete for their survival. In the current climate, government intervention in R&D worldwide has dwindled; hampering the level of innovation generated for public good. In fact, research facilities in many developing countries are poorly equipped, often limiting experiments to traditional and outdated research.

The diminishing role of public research institutes is perceived to have a major impact on the adoption of modern biotechnology in terms of introducing relevant products to those that need them most. Most of the field trials in the EU and the USA are conducted by private companies. An analysis of field-trial data from the USA shows that three crops account for 64% of all trials, of which 69% express herbicide- and pest-resistance traits. Of the trials conducted in the EU, 67% involve maize, sugar beet and rapeseed, and 71% of the novel gene categories presented herbicide- or pest-resistance traits. Less than 1% of all the trials in the EU and the USA are of plant varieties grown in tropical and subtropical climates, half of which have been conducted by the public sector.

Most of the public-sector research involving modern biotechnology in developing countries is still in the laboratory phase - none of the crops have progressed to marketable products. China, on the other hand, has approved the field-testing of over 500 GMOs to date and the commercial release of 50, including a prolonged shelf-life tomato, virus-resistant sweet pepper and vaccines for animal use. The experiences limiting the progression to commercializing research efforts range from: a lack of resources for meeting the high costs of regulatory requirements; lack of foresight, planning and business acumen for enabling the transition from research to a commercial product; and lack of capacity to negotiate patent licenses. Also, developments in modern biotechnology have occurred independently of the sustainable agricultural goals and priorities of the developing countries concerned. Furthermore, a needs assessment for a particular technology has often not been carried out before a research project is begun. Nevertheless, it is often argued that the commercialization of some products would encourage monocropping as national agricultural research has focused on a few crops, whereas rural communities tend to grow a wide range of crop species and plant varieties. The focus of international agricultural research centres is on plant production and protection, livestock production and health and food processing. With respect to food crops, research emphasis appears to be spread equally between cereals, root crops and legumes. However, within the cereals, research devoted to rice far outweighs research on maize and sorghum.

A large proportion of the total agricultural research activities of many developing countries are donor-funded. International research institutes, such as the Consultative Group on International Agricultural Research, depend on government grants and donations from philanthropic organizations for their survival, and yet investment in this sector has fallen in real terms. A significant fraction of the funding spent on international agricultural research institutes is used on activities covering a relatively large number of crops. The beneficiaries of such initiatives are a small number of countries with relatively advanced scientific capabilities. During the 1990s, developing countries as a group invested more in agricultural research than developed countries, even though the spending was unevenly distributed. In industrialized countries, private-sector investment in R&D far exceeds government spending on technology development, so that much of the public good previously entrusted to public research institutes is now privately owned. In comparison, private-sector investment in developing countries is around 1% of total global spending in this sector and developing countries invest less than 5% of the total private sector spending in biotechnology. Although the agricultural sector in developing countries is large and of significant importance to the domestic economy, spending in agricultural research does not match this level of activity. For instance, 80% of the food consumed in sub-Saharan Africa is obtained from domestic production.

Impact of Intellectual Property Rights on Research

Intellectual property rights have been relevant to agriculture since their inception but have gained importance with regard to research in developed countries in the past 20-30 years. In particular, IPRs have been used to protect and preserve the value of products produced by conventional methods, *e.g.* the trademark registration of food products. The rationale for IPRs is that they encourage the inventor to advertise the invention and disclose the new knowledge, while simultaneously holding the rights to protect the invention from competitors. Disseminating this information is thought to stimulate new ideas and further rounds of innovation and technological advancement. IPRs afford time-limited protection to artistic, scientific, technological or economic products, and can be protected by way of copyrights, trademarks, design patents, utility patents, plant patents, plant breeders' rights and trade-secret law. Of these mechanisms, patents are considered the most powerful tool of the IPR system. Patents play different roles in different technologies and sectors. Patent protection of biotechnology makes it a tool for technology transfer and securing new markets in a global economy. Without protection, new ideas and information are entirely in the public domain.

This can, in certain systems, lead to underinvestment in R&D or the withholding of knowledge. Plant variety protection provides less protection than patents in that generally it makes provision for farmers' rights, allowing them to use harvested seed, and includes an exemption for research use. Despite the increase in availability, new plant varieties continue to be inaccessible or inappropriate for poor farmers, and the rate of innovation remains largely unchanged in countries with a PVP system. Studies have indeed shown that in middle-income countries, the principal beneficiaries of PVPs are commercial farmers and the seed industry. PVP is seen as a system that protects small advances in plant breeding, while a patent regime is thought to lead to the protection of big leaps in technological achievements. Patent protection for products of modern biotechnology is important because they are expensive to develop and easy to copy. Even so, developing countries have limited capabilities to innovate in industrial fields such as modern biotechnology, and to effectively enforce IPRs. A significant number of developing countries have not established intellectual property regimes that cover plants.

This situation may thus discourage private-sector investment. With no assurance that they can recoup some profit on GM products, multinationals are unlikely to devote much attention to the challenges of developing countries unless seen in a development aid context or through public-private partnership. Although this situation impedes private-sector investment in developing countries, it also implies that the freedom to operate on products designated for local markets is not hindered. Exercising this freedom to operate is not a well-understood concept. For example, in the case of 'golden rice', where permission was required for the use of about 70 patents, the impression was that the patents were being relinquished in favour of the poor. In fact, most of the patents involved are not valid in the major rice-consuming countries. The technology donated for the development of virus-resistance in non-commercial potato varieties is free of patents relevant to Mexico, and the same holds true in the case of virus-resistant sweet potatoes in Kenya. Researchers are usually unaware of the proprietary status of the technologies they are using in their work. Nevertheless, the proliferation of broad patents is thought to impede the research capabilities of other interested parties. Some countries grant very broad patents conferring monopoly rights over large areas of research, thereby potentially threatening the other goal of intellectual property, namely the right to build upon the original invention. The prevailing patent rules have the potential to limit the accessibility of these technologies to public institutions and ultimately poor farmers.

Furthermore, the strengthening of IPRs is thought to restrict the flow of germplasm and inhibit the development of new plant varieties. This is because if and when researchers in public institutions do get permission to develop the technologies further, access is granted under licence agreements with restrictions on commercializing innovations. It is also argued that a stringent, multilateral IPR system will not benefit all countries equally. Indeed, the benefits will largely be influenced by the economic and technological levels of development in each country.

The United Kingdom Commission on Intellectual Property Rights, "the critical issue in respect of IPRs is perhaps not whether it promotes trade or foreign investment, but how it helps or hinders developing countries to gain access to technologies that are required for their development." During the 1990s, obtaining a patent application in the USA cost US$20,000 and twice as much in the EU. In general, PVP is cheaper; valued at one-tenth of the patent price. Moreover, the preparation of a food-safety dossier for a product derived from modern biotechnology, for example, is estimated to be around US$1 million.

These estimates cannot be compared with the regulatory costs in developing countries. The cost of regulation in developing countries does not encourage the commercialization of products of modern biotechnology developed by public-sector research institutes. In most cases, the regulatory costs far exceed the research costs. Many developing countries do not have the resources to match private-sector investment in modern biotechnology. In this new playing field, public institutions also need resources to deal with intellectual property rights to help compensate and increase the public benefit. Otherwise their involvement in R&D could be deterred by lack of funding. If public institutions are to use the techniques of modern biotechnology, then the use of IPRs as a framework for facilitating technology transfer must be emphasized more than its handling as a revenue-generating system. IPRs can, however, play a major role in clarifying the mechanisms for access to technology, and determine the downstream aspects of use and exploitation of genetic resources.

There are several ways in which public institutions and small companies in developing countries can gain access to patented genes and enabling technologies to overcome the current barriers to research. The first of these includes a measure of goodwill by multinationals to relinquish their rights to technologies for use by researchers in developing countries by adopting programmes of social responsibility as in the case of 'golden rice', a variety containing beta-carotene, virus-resistant sweet potatoes and a virus-resistant non-commercial potato variety in Mexico.

A different type of programme initiated in the USA has established an intellectual property clearing house to make information on the intellectual property owned by public research institutes, including universities, available to researchers around the world.

There are also suggestions that redesigning patent laws to narrow the type and scope of patent coverage ought to make more technologies accessible to public institutions. The thinking behind these suggestions is that applying a stronger standard for rejecting patent applications for inventions that are 'obvious' should deter the patenting of minor inventions. In addition, a law that requires an invention to be genuinely useful in theory should reduce the number of patent applications being submitted. At present, it is possible in some countries to submit patent applications for abstract concepts that potentially protect large areas of research and thereby exclude innovation by others. Another option that may be attractive to developing countries is the creation of collaborations that involve research institutes, universities and the private sector.

The nature of these collaborations is likely to be influenced by the level of expertise and resources within the national public research institutes. Where a solid knowledge base exists, the public partners may be in a position to develop or acquire a technology that could be transferred into locally adapted varieties. Smaller institutes are more likely to provide the genetic resources and a positive public image. It is believed that such alliances would benefit public institutions and private companies; offering them an opportunity to license and distribute the technology.

The most well-known public-private sector partnerships include organizations such as the International Service for the Acquisition of Agri-biotech Applications that negotiate access to private-sector technologies for the improvement of subsistence crops and/or the transfer of technology and know-how. Although several types of public-private sector alliances already exist, the two newly established initiatives worth mentioning are the Global Cassava Partnership and the African Agricultural Technology Foundation, launched in November 2002 by FAO and in March 2003 by the Rockefeller Foundation, respectively. The former is a partnership involving some of the world's leading experts in cassava research, working mainly in public institutions. The AATF intends to function as a clearing house of available technologies with the primary aim of improving food security and reducing the poverty situation of small farmers, by facilitating transfer and use of appropriate technologies.

Such licensing arrangements have been put to the test in other fields. However, as in the case of the AATF, a clearing house is required to acquire the necessary technologies and permit their further use for developing-country needs. The drawback may be a requirement to divide the commercial sector into subsistence, middle-income and commercial markets. This market division may not be easy to achieve as some large developing countries have both commercially important markets and subsistence farmers.

A briefing document commissioned by the United Nations Industrial Development Organization proposes six activities that build upon private-sector investment and enable biotechnology transfer:

- Enabling government policies;
- Access to up-to-date authoritative information;
- Regional brokering service to strengthen public-private partnerships;
- A regional biotechnology investment service;
- An international intellectual property escrow service; and
- Initiatives for risk-shifting.

Each of the proposals can be implemented as a stand-alone project or in combination, as best suits the national and/or regional situation.

Access to Genetic Resources

Historically, plant genetic resources were freely provided by developing countries to gene banks worldwide. The resources in question did not belong to a particular individual, and are often considered a common heritage of mankind. The application of modern biotechnology to genes that could be incorporated in genetic resources of importance in rural populations raises concerns in that the small-scale farmers may have originally supplied the genetic resources for improvement. Once privately owned, these resources may be unavailable to the people who have ensured their conservation for centuries.

Equally important is the aspect of access by researchers to genetic resources for further development on terms that recognize the contributions made by farmers to the conservation and sustainable utilization of these resources. At the international level, the importance of national ownership of such resources is duly recognized. The International Treaty on Plant Genetic Resources adopted at an FAO conference in 2001 provides the legal framework for dealing with the resources on which food security and sustainable agriculture depend. The Treaty gives a directive on the conservation and sustainable use of plant genetic resources for food and agriculture, making provision for the fair and equitable sharing of the benefits arising out of their use, in harmony with the principles of the Convention on Biological Diversity, but introducing the concept of farmers' rights. In the discussions on farmers' rights, the main issues of concern revolve around benefit-sharing and prior informed consent, and the protection of traditional knowledge from 'biopiracy'. This means access to genetic resources must be on mutually agreed terms to promote their use and emphasize their importance to development. A number of organizations are discussing the protection of traditional knowledge and folklore.

The Treaty establishes a multilateral system of facilitated access and benefit-sharing for key crops, emphasizing the interdependency of countries in terms of plant genetic resources for food and agriculture. Developing countries rich in genetic resources are encouraged to place germplasm in the MLS. The users of the material will sign a material transfer agreement, incorporating the conditions for access and benefit-sharing through a fund established under the Treaty. In return, the owners of the genetic resources will get a share of the benefits arising from their use and development in the way of information, technology transfer and capacity building. *Ex situ* material, collected before the entry into force of the CBD, does not come under the purview of the Convention, and would thus be dealt with under the Treaty.

So far, 35 food and 29 feed crops have been entered into the system. In principle, the genetic resources stored under this system are available for improvement to all interested researchers. Such wide-scale availability of germplasm has a potentially positive influence on access to improved technologies and nutritionally enhanced staple crops for the food-insecure. Genetic resources obtained from the MLS cannot be patented, even though it is not clear whether a gene isolated from such material may be protected or not.

9

Foods Containing Genetically Modified Materials

The first food products containing genetically modified material are now available on supermarket shelves throughout the world. Much public debate has arisen concerning the safety of such products and indeed the need for genetically modified foodstuffs in the well-stocked larders of the Western World. However, because genetic engineering offers such technical advantages to the food industry in the mass production of cheap processed food of predictable consistency and quality, there will be increased commercial pressure to broaden the range of genetically modified foodstuffs available in the marketplace.

Similarly, as our confidence in the safety of this new technology grows and our ability to apply genetic engineering to products which offer real benefits to the consumer (e.g. safer, more wholesome food) and in the development of 'functional' foods which may well play an important role in human health and disease prevention, consumer acceptance and demand for such products may grow. However, consumer confidence will rely on rigorous assessment of the various potential risks involved and appropriate safety testing. Microorganisms, particularly lactic acid bacteria, have been used in the production of fermented foods for millennia. Recent advances in genetic engineering allow, for the first time, accurate identification of microorganisms traditionally used in food fermentation and the design of novel strains with improved characteristics.

Age-old production problems such as instability of industrially important traits and failure of starter cultures due to bacteriophage attack may now be tackled at the molecular level. It has long been proposed that certain lactic acid bacteria or 'probiotics' contribute greatly to intestinal health and well-being. Similarly, genetically modified lactic acid bacteria have also been proposed for use as oral vaccines. Recent advances in molecular microbial ecology allow the scientific basis of such claims to be determined and genetic engineering will enable the design of probiotic bacteria with specific health-promoting properties. No food products containing live genetically modified microorganisms (GMMs) are available in the marketplace at present.

However, commercial pressure on the food industry and the consumer benefits promised by probiotic strains designed with scientifically proven health-promoting capabilities will encourage their use in fermented foods in the near future. Clearly the biosafety of such products must be rigorously investigated before they become commercially available, particularly in view of the public back-lash towards genetically modified plant material used in the food products in Europe.

Of particular concern when considering the release of live genetically modified microorganisms in food is the possibility of recombinant DNA transfer from genetically modified organisms (GMOs) to members of the human gut microflora. Transfer of DNA between bacteria occurs naturally in the environment and offers prokaryotes a unique means of evolution and adaptation in response to changing environmental conditions.

Interest in DNA transfer between bacteria in the mammalian gastrointestinal tract stems from three main areas:

- The possibility of DNA transfers from GMOs which may be ingested in food to members of the human gut microflora.
- The spread of antibiotic resistance amongst bacteria as a result of gene transfer.
- The emergence of novel human pathogens as a result of transfer of virulence factors or antibiotic resistance determinants between bacteria.

The focus of this review is to discuss the ability of bacteria to undergo DNA transfer in the human gastrointestinal tract with particular reference to the risks posed by genetically modified microorganisms, which may be used in the production of fermented foods such as bread, beer, cheese and yoghurt. We will look at the existing procedures available to monitor DNA transfer in the human gut microflora and investigate some of the factors governing frequencies of such transfer events.

These studies will provide information relevant to our understanding of the possibility of the transfer of marker genes used in the construction of genetically modified crops to the bacteria present in the mammalian gastrointestinal tract. First let us look at the mechanisms of DNA transfer available to bacteria in natural environments.

TRANSFORMATION

Transformation is the process by which a naked piece of DNA from the environment binds to the surface of a competent bacterial cell and is taken up by the bacterium. DNA may then be incorporated into the host genome, depending on the recombinational abilities of the host and the 'foreign' DNA. The ability to translocate DNA across the cell boundary is called competence.

Competence is a specific physiological state of a bacterial cell (genetically encoded in some cases, e.g. Bacillus subtilis), which occurs transiently and is restricted to certain stages of the growth cycle. Natural competence has been observed in bacteria from a variety of genera, including Haemophilus, Neisseria, Streptococcus, and Bacillus, Acinetobacter and Pseudomonas spp. and Helicobacter pylori.

Recalcitrant species, such as Escherichia coli, may be rendered competent using a variety of chemical, enzymatic and physical procedures, e.g. CaCl2 treatment and electroporation. Such physiochemical conditions may sometimes be prevalent in the local environment of recalcitrant bacteria. For example, Ca2+ concentrations in drinking water sometimes approach the levels used in vitro to induce a state of competence in E. coli.

Competence is not the only bacterial-encoded parameter shown to play a role in natural transformation. In some cases, requirements for specific lengths of DNA, DNA states (double or single stranded) and the presence of specific DNA sequences have been observed. Competent Bacillus and Streptococcus spp. may take up any piece of DNA but only homologous DNA will be maintained in the bacterial genome.

Haemophilus spp., on the other hand, requires the presence of specific 11-bp sequences before DNA uptake occurs. These 11-bp recognition sequences occur at a number of locations on the Haemophilus genome. It has also been reported that a diffusible factor may play a role in the induction of competence in Streptococcus spp. Here, induction of competence was found to be dependent on cell density and the pH of the surrounding environment.

Factors Affecting Transformation in Natural Environments

The presence of naked DNA has been demonstrated in a number of natural environments. Naked DNA may arise in a given habitat via a number of routes. Cell lysis as a result of cell death or the activities of bacteriophage will release bacterial DNA and DNA may be released from actively growing bacteria during certain stages of the growth cycle.

The uptake of DNA from the environment will not only depend on a state of competence in the bacterium but also on the persistence of naked DNA in a given environment. In this respect, different microhabitats will vary greatly in their abilities to either protect naked DNA by the presence of favourable salt concentrations or absorption onto solid supports (e.g. the surface of soil particles or food particles in the gut) or to degrade DNA, for example, by active nucleases present in the microenvironment.

Despite the longevity of DNA in the soil and the presence of bacteria with known competence for transformation, gene transfer from plant to soil bacteria appears to be an extremely rare event: for example, transfer of an ampicillin resistance gene from a transgenic potato line to the plant pathogenic bacterium Erwinia chrysanthemi was at a calculated frequency of 2 × 10-17 and from transgenic plants to the soil bacterium Acinetobacter calcoacetinus at a frequency lower than 10-13. However, the uptake and integration of transgenic plant DNA via natural transformation has been shown for Acinetobacter sp., strain BD413 by in vitro marker rescue using DNA from various transgenic plants containing the bacterial kanamycin-resistance gene nptII. Naked DNA enters the human gastrointestinal tract from a number of sources. These include ingested food and foreign microorganisms as well as members of the human gut microflora.

Very little is known about the ability of naked DNA to persist in the gut and evade the activities of mammalian and bacterial nucleases. Factors which may affect the persistence of naked DNA, and thus the incidence of transformation in the human gastrointestinal tract, include pH, salt concentrations, cell densities, local nuclease activities and protection afforded by absorption onto surfaces such as food particles or mucosal surfaces.

Such factors would act at the level of the microhabitat and as such would vary greatly even within specific regions of the gut. It has been shown that the CRY1A protein and DNA from genetically modified maize are digested by simulated gastric juices in vitro. However using an in vitro model of the intestinal tract, van der Vossen *et al.* showed that 6% of transgenic tomato DNA survived the stomach and small intestine and concluded that the presence of raw mashed tomato helped to preserve the DNA. Free chromosomal DNA of Bacillus subtilis persists for weeks in milk and dairy produce. Such bacteria develop natural competence and can be transformed with free chromosomal or plasmid DNA in such produce. Schubbert *et al.* showed that the wall of the gastrointestinal tract is exposed to a variety of DNA fragments and remains exposed to DNA fragments of dietary origin for hours after ingestion of the food.

The authors found that upon feeding mice 50 μg M13mp18 DNA, approximately 95% of ingested DNA was lost during passage through the stomach. However, phage DNA could be detected by PCR and fluorescent in situ hybridization in peripheral leukocytes, spleen and liver cells as well as the contents of mouse small intestine, caecum, large intestine and faeces.

Ingested phage DNA was detected for up to 18 hours after ingestion in caecal contents, for up to 8 hours in DNA from the peripheral blood cells and for up to 24 hours but not 48 hours in DNA from spleen and liver. Mercer *et al.* monitored the survival of recombinant plasmid DNA (pVACMC1) in fresh human saliva. The fraction of naked DNA remaining amplifiable in saliva ranged from between 40 and 65% after 10 minutes and 6-25% after 60 minutes. Amplifiable plasmid DNA was still present after 24 hours incubation in fresh saliva. The authors also found that plasmid DNA, which had been exposed to degradation by saliva, was capable of transforming naturally competent Streptococcus gordonii DL1 in filtered saliva. Transformation activity decreased rapidly with the extent of plasmid DNA degradation.

Such studies suggest that DNA released from food or bacteria ingested with food may undergo transformation in not only the oral cavity but other regions of the human gastrointestinal tract. Clearly, much more research is needed on the ability of naked DNA to persist in the lower regions of the human gastrointestinal tract and the extent to which transformation contributes to DNA transfer in the human gut microflora.

TRANSDUCTION

Transduction is the mechanism by which DNA may be transferred between bacteria by bacteriophage. In essence, the bacteriophage acts as microbial couriers, picking up DNA from one bacterial chromosome and delivering the heterologous DNA to another bacterial chromosome.

Where degradation of naked DNA is one of the chief factors limiting DNA transfer by bacterial transformation in natural environments, bacteriophage protects DNA in natural environments via their proteinaceous capsid. Bacteriophages, on the whole, contribute to gene transfer between bacteria by two main mechanisms, namely specialized and generalized transduction.

Specialized Transduction

Specialized transduction involves the incorporation of a lysogenic phage into the bacterial chromosome. Lysogeny is favoured by conditions of environmental stress such as the limitation of nutrients and probably aids the survival of the phage and host bacterium. Upon excision from the chromosome, elements of host DNA adjacent to the prophage DNA may become excised along with the phage DNA. The host DNA may then become incorporated into the bacteriophage genome and packaged along with the phage DNA. Thus, when infection of a recipient bacterium occurs the heterologous DNA may become integrated into the recipient's chromosome during lysogeny.

Generalized Transduction

Generalized transduction on the other hand, occurs when host DNA is packaged into phage particles instead of the phage genome. This mechanism of transduction has been observed in the lytic phage, P1 and P2 of the enterobacteraceae, for example. Upon transduction to a novel recipient, the transferred DNA may either be incorporated into the host chromosome via recombination or where it possesses the means of self-replication, it may replicate autosomally.

Factors Limiting DNA Transfer by Transduction

Transduction is greatly dependent on the host range of the transducing bacteriophage, which is generally narrow, since bacteriophage infection is dependent on the phage recognizing specific receptor sites on the bacterial cell surface. Some bacteriophage that are able to mediate transduction between different species of bacteria have been described, e.g. P1 and Mu. The frequency of transduction in nature may be much higher than previously recognized, since the number of bacteriophage particles in many environments appears to be much higher than first thought. Whether the transferred DNA is maintained in the new host is greatly dependent on the ability of the bacteriophage to insert itself and the heterologous DNA into the host genome and evade the bacterial restriction modification system. Bacterial restriction enzymes may recognize specific sequences on heterologous or phage DNA. It has been proposed that restriction modification systems may reduce infection of unmodified phage DNA by 2-3 orders of magnitude. The amount of DNA that may be transferred by transduction is also limited, being about equivalent in size to the phage genome itself.

Transduction in the Human Gastrointestinal Tract

Despite the fact that the bacteriophage are ubiquitous members of natural microbial ecosystems, little is known about their distribution or activity in the human gut microflora. To date no reports of transduction in the human gastrointestinal tract have been presented. However, where sufficient effort has been made to isolate bacteriophage from natural environments and examine their activity in situ, they have been found to play a significant role in microbial ecology.

The fact that about 90% of all bacteriophage isolated from natural environments are temperate suggests that transduction may be more important in microbial genetic plasticity than previously appreciated. Thus specifically designed studies involving in vitro or in vivo models of the human gastrointestinal microflora may well provide evidence of transduction in this ecosystem and elucidate some of the ecological factors governing transduction.

Bacteriophage specific for major groups of bacteria present in the gut microflora, e.g. Bacteroides spp., bifido-bacteria, lactobacilli, methanogens, clostridia, enterobacteriaceae, streptococci and staphylococci have been identified. However, transduction has not been observed in many of these bacterial groups e.g. Bacteroides spp. or Clostridium spp.

Bacteriophage of the lactic acid bacteria play a major role in the dairy industry both as destructive agents, causing the failure of starter cultures in cheese and yoghurt production, and as valuable genetic tools in the development of genetically altered industrial strains. The variety of bacteriophage associated with the lactobacilli suggests that transduction may be a significant means of gene transfer within these genera. Tohyama *et al.* demonstrated that the Lactobacillus salivarius temperate phage PLS-1 mediates generalized transduction of auxotrophic markers (lysine, proline and serine) and lactose metabolism at frequencies of 10-7 to 10-8 transductants per CFU in vitro. Later, Raya *et al.* showed that phage ADH replicates in a lytic cycle, establishes lysogeny, confers superinfection immunity on the host and mediates plasmid DNA transduction in L. acidophilus ADH. It was also observed that plaque formation on cell lawns of L. acidophilus NCK102 was pH dependent, with the optimal pH for plaque formation being pH 5.5. Transfer of antibiotic resistance determinants have been reported between strains of Desulfovibrio desulfricans via phage Dd1 mediated transduction at frequencies of 10-5 to 10-6 transductants per recipient.

Transduction has also been observed in methanogens, although not strains found in the human gut microflora. Despite demonstrations that transduction does occur under labouratory conditions, little is known about the significance of transduction-mediated gene transfer in the natural and animal-associated environments.

Important questions remain unanswered regarding the prevalence and survival of bacteriophage in different environments, the limitations imposed on transduction by the host specificity of the transducing bacteriophage and the frequencies at which transduction occurs during phage replication.

GENE TRANSFER FROM GENETICALLY MODIFIED MICROORGANISMS

The Consultation noted that there are well-known mechanisms of transfer of genetic material between microorganisms, such as transduction and conjugation. Transformation of naked DNA into microorganisms in the GI tract has not been conclusively demonstrated. The probability of gene transfer in the GI tract has to be assessed in the light of the nature of the genetically modified organism and the characteristics of the gene construct. Possible consequences of a transfer event should be assessed based on the function and specificity of the transgene. The likelihood of maintenance of the transferred gene in a recipient microorganism increases if the gene confers to the microorganism a selective advantage. Factors that may enhance the selective advantage over other organisms or the colonization ability include: phage resistance, virulence, adherence, substrate utilization or production of bacterial antibiotics. If the transferred gene is not expected to enhance any of the survival characteristics of the recipient gastrointestinal microorganism, no further safety assessment concerning these characteristics would be required. If the function of the gene suggests that survival of the recipient organism would be enhanced, the possible health consequences need to be assessed, based on the function and specificity of the gene.

The Consultation affirmed the recommendations from the 1990 FAO/WHO joint consultation regarding genetically modified microorganism including:

- That vectors should be modified so as to minimize the likelihood of transfer to other microbes; and,
- Selectable marker genes that encode resistance to clinically useful antibiotics should not be used in microbes intended to be present as living organisms in food.

Food components obtained from microbes that encode such antibiotic resistance marker genes should be demonstrated to be free of viable cells and genetic material that could encode resistance to antibiotics. The Consultation was not aware of any reports of transfer of genes from animal, plant or microbial origin into epithelial cells except for genes from infectious agents, such as viral DNA. However, even if such transfer were to occur, the transformed epithelial cells would not be maintained in the GI tract due to continuous replacement of these cells.

PATHOGENICITY OF MICROORGANISMS

The Consultation reviewed the 1990 report discussion of the issue of pathogenicity related to genetically modified microorganisms, and agreed that no new issues have arisen since that consultation. To summarize, microorganisms intended for use as food or in food processing should be derived from organisms that are known, or have been shown by appropriate tests in animals, to be free of traits that confer pathogenicity. Furthermore it was stated that assessment of viable genetically modified organisms as part of a food must also take into consideration characteristics that determine their survival, growth and colonizing potential in the GI tract, including the capability to undergo transformation, transduction and conjugation, and to exchange plasmids and phages. In this regard, a general principle was elaborated that design should be directed towards minimizing intrinsic traits in microbes that allow them to transfer genetic information to other microorganisms.

GENETICALLY MODIFIED ANIMALS

The 1990 consultation reviewed the safety assessment of genetically modified animals and foods derived from them and concluded inter alia that:

- "Mammals are important indicators of their own safety, since adverse consequences of introduced genetic material will generally be reflected in the growth, development and reproductive capacity of the animal. The principle that healthy mammals only should enter the food supply is of itself a method of ensuring the safety of foods derived from animals. Primarily because some fish and invertebrates are known to produce toxins, the healthy animal principle does not provide the same degree of assurance that food derived from such animals is safe and should be used with caution in determining the need for additional safety assessment".

The OECD report on "Safety Evaluation of Foods Derived by Modern Biotechnology: Concepts and Principles", focused its attention on new foods of terrestrial origin and concluded that, "In general, foods from new strains of mammals and birds that appear to be in good health have proven to be as safe as the animal breeds from which they were derived." Subsequently OECD convened a workshop on "Aquatic Biotechnology and Food Safety" at which was discussed the notion that when an animal appears healthy, this is an indication that the animal is safe to eat. It was recognized that the apparent good health of aquatic food organisms *per se* is not a useful indicator of food safety, because many such species are known to contain either exogenously or endogenously derived compounds that are toxic to humans. Individuals of such species frequently appear to be in good health because they are resistant, in some degree, to these toxins. However, if the notion of healthy appearance is applied in this sense, and if it is applied in conjunction with other attributes to assess safety, then it can still have utility when applied to food and food components derived from aquatic animals.

In general, the OECD workshop on Aquatic Biotechnology and Food Safety considered it unlikely that techniques of modern biotechnology will increase the risk to human health if applied to aquatic organisms containing toxins. It is possible that modern breeding techniques could affect the metabolism and properties of such toxins, perhaps in ways that modify their effects. In such circumstances, however, the safety assessment of such organisms will depend on knowledge of, and on techniques to detect, toxins. Consequently, the fact that some aquatic food organisms might contain compounds toxic to humans does not reduce the value of the application of the concept of substantial equivalence.

The OECD workshop also concluded that no issue could be identified which reduced or invalidated the application of the principle of substantial equivalence to food or food components derived from modern aquatic biotechnology. However, it was recognized that in some instances there may be a lack of appropriate data from the conventional species. This lack of data could lead to difficulties when making comparisons with the new food or food component. This problem arises, in part, because there is less familiarity with most aquatic organisms in food production when compared with terrestrial food animals and plants. If animals are genetically modified to improve their resistance to bacteria and viruses that also represent a human health concern, appropriate hygiene measures should be taken to ensure that there are no food safety risks to consumers of the animal products.

FOOD ORGANISMS EXPRESSING PHARMACEUTICALS OR INDUSTRIAL CHEMICALS

Genetic modification has considerable potential to enable the production of pharmaceuticals or industrial chemicals in varieties of organisms, that are also used as sources of food. The Consultation recognised that, generally, the genetically modified organism would not be used as food without prior removal of the pharmaceutical or industrial chemical. The Consultation agreed that the safety assessment of pharmaceuticals and industrial chemicals, as such, was outside its remit. In situations in which the genetically modified organism or its products are used in food, the Consultation agreed that the concept of substantial equivalence, as developed elsewhere in this report, could be used for the safety assessment of the food. Some such foods could be substantially equivalent to existing foods, apart from well defined differences whilst others might be substantially equivalent to their conventional counterparts.

There might also be situations in which the food would not be substantially equivalent to an existing counterpart. In addition to concerns about food safety, the Consultation recognised that the genetic modification of food organisms to produce pharmaceuticals or industrial chemicals may raise ethical and control issues that were outside its remit because the issues were unrelated to food safety. The ethical issues relate to the scope for administering treatments to consumers without their knowledge. The Consultation agreed that this issue should be brought to the attention of WHO.

DATABASES

To facilitate the compositional comparisons necessary to establish substantial equivalence, it may be useful to use and even generate international databases containing validated data on the nutrient, allergen and, especially, toxicant composition of commonly used food organisms. If the genetically modified organism is being compared directly to its parent then the data will be developed when the modified organism and the parental organism are grown and analysed under a limited number of selected environments that are representative of the conditions under which the modified organism will be used commercially. Comparison of genetically modified plants with other commercial varieties will typically focus on data generated within these varieties grown within similar geographical regions and in which the new variety is intended to be grown commercially. Where compositional analyses of key nutrients and toxicants are required for the registration of a new plant variety, these data would be updated periodically for current commercial varieties.

The published literature on these parametres would also serve as a source for this information. It is important that the reference ranges represent reasonably current information since the ranges will probably change over time. In the case of plants, relevant information could be obtained from the international centres of the Consultative Group on International Agricultural Research which holds the worldwide mandate for the conservation and use of genetic resources. These include the Centres for Genetic Resources Conservation and Breeding Research on specific crops; *e.g.* the International Maize and Wheat Improvement Centre, the International Rice Research Institute, the International Potato Centre and the International Plant Genetic Resources Institute.

The FAO and the World Food Programme provide other sources of information on food composition. Databases on microorganisms, mainly in the form of type culture collections, are in existence but not all are suited for the purpose of establishment of substantial equivalence. The Consultation acknowledged that molecular databases are available and commonly used to identify genes/proteins of similar structure and/or function. These databases are also used to compare amino acid sequence homology of an encoded protein to known protein toxins or allergens. These databases should continue to expand as new genes/proteins are isolated and characterized. The Consultation pointed to the need to develop and expand databases with valid data on the content and ranges of nutrients, toxicants and allergens in food organisms used throughout the world.

THE APPLICATION OF RDNA TECHNOLOGY IN DEVELOPING COUNTRIES

Recombinant DNA technology has broad application in developing countries and has the potential for very positive impact on their economies, which are frequently agriculturally based. In this context the view has been expressed that rDNA technology might be of greater importance for developing countries than for industrialized countries. In particular, developing countries look on rDNA technology as a means of addressing the need to produce sufficient quantities of nutritionally adequate and safe food for their growing populations.

The benefits of this technology are likely to impact directly on people at the production level as this technology is extremely easy to transfer, being "packaged in a seed". However, in order for the entire global population to fully benefit from rDNA technology, the safety assessment of food derived from genetically modified organisms requires trained manpower, up-to-date legislation and a food control system for its enforcement. This applies equally in all the countries of the world.

Food safety issues are not bound by national borders, and it is therefore important that countries that have inadequate resources for assessing rDNA technology and products derived from it, make special efforts to obtain these resources. Moreover, since globalization interconnects raw material production to processing and consumers of all regions of the world, it is imperative that proper safety assessment of foods and food components produced by genetic modification, be practised world wide.

MICROBIAL DETERIORATION OF FOOD COMPONENTS

The type and extent of microbial colonization of a food only partly affects its ultimate deterioration, because the biochemical activities of the microbial community structure at the time of the onset of spoilage are also decisive. Organoleptic deterioration may, however, occur before any marked chemical changes take place in the food. This is because some odiferous metabolites can be detected organoleptically at very low levels. Less than 1 ppm dimethyl sulphide or methyl mercaptan is sufficient to cause off-odours. Even at the maximum cell concentration usually achieved, metabolising at the optimum rate would only produce about 2 ml g-1 h-1 of carbon dioxide. At lower temperatures this rate would be much less. Conversely, high levels of microbes may be present in a food that shows no obvious organoleptic change. The growth of microbes in foods inevitably causes chemical changes. Bacteria, the predominating organisms in the microbial ecology of most foods, are extremely small: a rod of 2 × 0.8 μm has a volume of about 10-12 cm^3. Although they have a high metabolic potential per cell, large numbers of bacteria are required before they can cause measurable chemical changes.

MICROBIAL METABOLITES

Biological as well as fabricated food structures will possess receptors to which microorganisms can absorb. The resulting colonization of such structures may occur in a stratified way, leading to relatively high local concentrations of microbial metabolites. The metabolites formed by a given spoilage association will once again depend on the prevailing intrinsic, extrinsic and implicit conditions.

These include the limiting factors influencing:

- The type of spoilage, determined by the relative amounts of metabolites formed; and
- The rate at which these metabolites are produced during storage and distribution of the food.

The latter is mostly expressed as the time to spoilage, as detected by sensory evaluations—odour, colour, structure and taste. The microbial metabolites depend not only on the storage conditions but also on other environmental factors such as aeration, glucose and lactate availability, and pH.

Carbohydrates

Carbohydrates, if available, usually are preferred by microorganisms to other energy-yielding foods. The carbohydrates are divided into monosaccharides, disaccharides, and polysaccharides. The monosaccharides are polyhydroxy aldehydes or polyhydroxy ketones. For utilisation, bacteria first need to break down complex carbohydrates such as starch into their constituent monosaccharides. The random splitting of glycosidic bonds results in softening and liquefaction. Several bacteria possess an extracellular enzyme, diastase or amylase, which hydrolyses the starch.

The starch is then converted either directly to glucose or via intermediates such as maltose. Although flavobacteria do not degrade lignin and cellulose, it is possible that these organisms are involved in the breakdown of various proteins and carbohydrates. Glucose is the main carbohydrate used as a carbon and energy source.

The breakdown of this monosaccharide can proceed by several pathways. In aerobic respiration the glucose metabolite, pyruvate is converted into carbon dioxide and water by means of the tricarboxylic acid cycle, Krebs cycle, or citric acid cycle. To enter the system, the pyruvate is converted to acetate activated by coenzyme A. Only the aerobic and some facultatively anaerobic microorganisms possess an intact TCA cycle. The pyruvic acid can be decarboxylated to form acetaldehyde and CO_2. The acetaldehyde can remain or be reduced to ethyl alcohol, oxidized to acetic acid, or condensed to form acetoin or acetylmethylcarbinol. The AMC can be oxidized to diacetyl, which has a butter flavour, or reduced to 2,3-butanediol. Pyruvate can be aminated to form alanine. Boers *et al.* observed that the glucose concentration had decreased to a low level at the first signs of spoilage. It has been concluded also that glucose limitations caused a switch from a saccharolytic to an amino acid degrading metabolism in at least some bacterial species.

Foods with high levels of carbohydrates are preferentially colonized by glycolytic organisms and tend to ferment rather than putrefy. This leads to the production of acids and is accompanied by a reduction in pH. The lactate occurring in flesh foods due to post mortem glycolysis can often be differentiated by its optical rotation from lactic acid formed by microorganisms; this increases its reliability as an index of spoilage. However, in some instances lactic acid may be dissimilated and acetic acid may be a better indicator of microbial colonization and metabolism.

Fats

The principle lipids in foods are fats. Fats are esters of glycerol and fatty acids and are called glycerides, in the ratio of one molecule of glycerol to three molecules of fatty acids. A pure fat is not attacked by microorganisms, since there must be a nutrient-containing aqueous phase in which the organism can grow. Lipase, an enzyme that hydrolyses fats to free fatty acids and glycerol, is present in many kinds of foods.

Because milk contains an appreciable amount of this enzyme, milk fat often undergoes lipase-catalyzed hydrolysis with the production of free fatty acids, diglycerides, monoglycerides, and in extreme cases, free glycerol. Short-chain water-soluble fatty acids cause obnoxious rancid flavours in milk. Lipolysis in foods followed by ß-oxidation produce ketones, which always result in off-flavours. The oxidative deterioration of fats involves the reaction of unsaturated fatty acids with oxygen to yield hydroperoxides. The hydroperoxides are not flavour compounds, but readily decompose to carbonyl compounds resulting in off-flavours or -odours. The carbonyl compounds are mixtures of saturated and unsaturated aldehydes and produces ketones.

Proteins

Microorganisms, through their proteolytic enzymes, break down protein into simpler substances. The breakdown usually follows the following pattern:

Protein →Peptones →Polypeptides →Peptides →Amino acids →Ammonia → Elemental nitrogen.

Proteinases catalyze the hydrolysis of proteins to peptides, which may impart a bitter taste to foods. Peptidases catalyse the hydrolysis of polypeptides to simpler peptides and finally to amino acids. The latter impart flavours, desirable or undesirable, to some foods; e.g., amino acids contribute to the flavour of ripened cheeses.

The products that are formed depend upon the type of microorganism; the types of amino acids; temperature; the amount of available oxygen; and the types of inhibitors that might be present. Decomposition of protein by aerobic organisms is called decay. Proteins containing amino acids with sulphur, such as cystine and methionine, can be broken down with no unpleasant odour because the end products are completely oxidized and stabilized.

Sulphur compounds, however, are often associated with "putrid" odours. The metabolites produced by microorganisms in proteinaceous foods such as meat include ammonia, ethanol, lactate, acetate, indole and acetoin, with smaller quantities of higher fatty acids, amines and ethyl esters of the lower fatty acids, sulphides, hydrogen sulphide and mercaptans. Most of the esters, amines, ammonia and sulphur compounds are produced from amino acids.

There is no significant degradation of protein proper until spoilage has progressed to obvious deterioration. Owing to production of amines and ammonia, the pH of proteinaceous foods tends to rise as spoilage progresses. An increase in the pH of a protein food indicates protein degradation, just as a decrease in pH results from the fermentation of carbohydrates.

VOLATILE COMPOUNDS

In fresh products, such as fruit, vegetables and milk, flavour components are very abundant. Chang stated that while the odour of some foods may be accounted for by single key compounds, most food odours are the result of complex mixtures. Dainty *et al.* stated that the variability of individual chemicals found in the aroma of spoiled samples was not significant. A better understanding of the complexity can be gained if the volatile compounds are grouped into classes: sulphur compounds, ketones, esters, aromatic hydrocarbons, aliphatic hydrocarbons, aldehydes and alcohols, but just as to Fedele *et al.* found that not all of these compounds have significant effects on the overall odours. Examination of volatile compound profiles indicated there were at least 3 requirements for development of putrid odours.

These requirements are that:

- The total volatile compound peak area must be appreciably high,
- With exception of aliphatic hydrocarbons, the sulphur compounds must be the major constituents of the profile, and
- Large quantities of other classes, if present, may modify the effects of the sulphur compounds.

The determination of total "volatile bases", which include ammonia, trimethylamine and other compounds, correlates well with organoleptic judgment in a number of species of fish. Stutz *et al.* found that the concentration of four of the volatile compounds, acetone, methyl ethyl ketone, dimethyl sulphide and dimethyl disulphide increased continuously during storage of minced meat stored aerobically at 5, 10, or 20°C. Hydrogen sulphide and ammonia are formed as a result of the conversion of cysteine to pyruvate by the enzyme cysteine desulphydrase. Acetoin is the major volatile compound produced on raw and cooked meats in O_2-containing atmospheres. Overton and Manura milk samples were found to contain numerous straight and branched chain hydrocarbons, aldehydes, alcohols, ketones, fatty acids, esters, phenolic compounds and lactones.

BIOGENIC AMINES

Biogenic amines are basic nitrogenous compounds formed mainly by decarboxylation of amino acids or by amination and transamination of aldehydes and ketones. Biogenic amines in food and beverages are formed by the enzymes of raw material or are generated by microbial decarboxylation of amino acids, but it has been found that some of the aliphatic amines can be formed *in vivo* by amination from corresponding aldehydes. Koessler *et al.* proposed that biogenic amine formation is a protective mechanism for bacteria against acidic environments. The production of amines requires the availability of free amino acids and appropriate status of environmental factors such as pH and temperature.

The pre-requisites for biogenic amine formation by microorganisms are:

- Availability of free amino acids, but not always leading to amine production;
- Presence of decarboxy lase-positive microorganisms; and
- Conditions that allow bacterial growth, decarboxylase synthesis and decarboxylase activity.

Biogenic amines are present in a wide range of food products including fish products, meat products, dairy products, wine, beer, vegetables, fruits, nuts and chocolate. Virtually all foods that contain proteins or free amino acids and are subject to conditions enabling microbial or biochemical activity, are conducive to the production of biogenic amines. The total amount of the different amines formed strongly depends on the nature of the food and the microorganisms present. Different biogenic amines have been detected in fish such as mackerel, herring, tuna, and sardines.

Other amines, such as trimethylamine and dimethylamine are present in fish and fish products at levels depending on the fish freshness. Bacterial-produced histamine has also been found in dairy products and vegetables. Amines are also important because of their role in causing spoilage of dairy products by producing typical off-flavours and putrid odours. Putrescine, cadaverine, histamine, tyramine, spermine and spermidine were found to be present in minced pork, beef and poultry stored at chill temperatures. Histamine has been recognized as the causative agent of scombroid poisoning as well as nausea, vomiting, gastrointestinal distress and headache, whereas tyramine has been related to food-induced migraines and hypertensive crisis in patients under antidepres sive treatment with monoamine oxidase inhibitor drugs. Secondary amines such as putrescine and cadaverine can react with nitrite to form heterocyclic carcinogenic nitrosamines, nitrosopyrolidine and nitrosopiperidine.

The levels reported for histamine and its potentiators in food would not be expected to pose any problem if normal amounts were consumed. Sandler *et al.* reported that 3 mg of phenylethylamine causes migraine headaches in susceptible individuals, while 6 mg total tyramine intake was reported to be a dangerous dose for patients receiving monoamine oxidase inhibitors. The level of 1,000 mg kg-1 is considered dangerous for health. This level is calculated on the basis of foodborne histamine intoxications related to amine concentration in food. The European Community has recently proposed that the average content of histamine should not exceed 10-20 mg/100 gm of fish.

GENETICAL MODIFIED ORGANISM-DERIVED FOODS

The concept of substantial equivalence was originally developed through discussions at the Organisation for Economic Co-operation and Development, though, to a large extent, these discussions built on previous work done by the World Health Organization (WHO) and the Food and Agriculture Organization. In 1991, a Joint FAO/WHO Consultation had concluded that the evaluation of a food derived through modern biotechnology should consider both food safety and nutritional value using similar conventional food products as a standard and taking into account the processing of the food and its intended use.

In the 1980s, the OECD established a Group of National Experts on Safety in Biotechnology (GNE), which continued to work through to 1993. As an intergovernmental organization, OECD's GNE was attended by delegates nominated by the governments of the OECD member countries, who were, for the most part, representatives of those agencies and ministries with a responsibility for safety in biotechnology. As a result of the work of the GNE, the OECD published a number of important documents relevant to the safety assessment of biotechnology-derived products during this period, which deal with a range of issues, including safety considerations for industrial, agricultural as well as environmental applications of organisms derived by recombinant DNA techniques.

It was recognized at this stage that the safety assessment of an organism derived through recombinant DNA techniques would rely heavily on the knowledge of its parental organism, as well as on an analysis of how the new organism appears to differ from the parent. OECD recommended that the considerable data on environmental and human health effects of living organisms that exists should be used to guide the risk assessment.

By the early 1990s, there had already been large numbers of small-scale field trials of new crop varieties derived through recombinant DNA techniques, and it became clear that new foods derived from these varieties would be marketed during the 1990s. In order to proactively address food safety issues related to these novel foods, in 1990, the GNE established a Working Group on Food Safety and Biotechnology comprising experts, mainly from the ministries and agencies responsible for food safety issues in OECD member countries. The main objective of the Working Group was to elabourate scientific principles and concepts to be used when evaluating the safety of new foods and food components of terrestrial microbial, plant or animal origin.

The Working Group did not consider the safety assessment of food additives, contaminants, processing aids or packaging materials. Nor did it consider environmental safety issues which had been (or were being) addressed by other groups of the GNE. Early in the life of the Working Group, an important recognition was that traditionally, the safety of food for human consumption had been based on the reasonable certainty that no harm will result from intended uses under the anticipated conditions of consumption. Foods prepared and used in traditional ways have usually been considered safe on the basis of long-term experience, even though they may have contained natural toxicants or anti-nutritional substances.

Normally, new varieties of foods or crops have not been subjected to traditional toxicological testing. In fact, where toxicological testing has been applied to whole foods, the results have often been difficult to interpret. Although the OECD's Working Group recognized that modern biotechnology might extend the scope of genetic changes that can be made - and might even broaden the range of the possible sources of foods - it was recognized that this would not inherently lead to foods that are less safe than those developed through conventional techniques.

The Working Group therefore indicated that the evaluation of foods or food components derived through modern biotechnology does not require a fundamental change in established principles, nor does it require a different standard of safety. Realizing the importance of using examples of new foods to identify and demonstrate the applicability of the proposed scientific principles, the Working Group organized a number of meetings and intergovernmental consultations which included case study presentations of novel foods such as enzymes, genetically modified bakers' yeast, mycoprotein and a number of genetically modified varieties of crop species.

Although these case studies were not intended to be formal safety evaluations they were crucial in illustrating important concepts in the safety assessment of novel foods. Therefore, it was through their application in real examples that the concepts and principles were identified. The Working Group proposed a scientific approach to the evaluation of foods derived through modern biotechnology, which is based on a comparison with traditional foods that have a safe history of use. One of the main concepts developed was that of substantial equivalence, and in describing this concept, the Group stressed that this was a principle that had been used in the past (perhaps intuitively) even if it had not been articulated as such.

The concept of substantial equivalence is described as embodying the idea that existing organisms used as food, or as a source of food, can be used as the basis for comparison when assessing the safety of human consumption of a food or food component that has been modified or is new. As previously indicated, in elabourating the concept of substantial equivalence, the OECD Working Group noted that food safety is considered as a reasonable certainty, that no harm will result from intended uses under the anticipated conditions of consumption and that the most practical approach to the determination of safety is to consider whether a genetically modified organism (GMO)-derived food is comparable to an analogous conventional food product.

The concept of substantial equivalence therefore relies on the existing history of safe food use of a conventional food product as a useful consideration in framing the safety assessment of a GMO-derived food by permitting the identification of similarities and differences which can be considered in the assessment. In 1996, participants at an expert FAO/WHO consultation recommended that safety assessment based upon the concept of substantial equivalence be applied in establishing the safety of foods and food components derived from genetically modified organisms.

Assessing substantial equivalence was recognized as not being a safety assessment per se, but a process which establishes that the characteristics and composition of the new GMO-derived food are comparable to those of a familiar, conventional food which has a history of safe consumption. A Joint FAO/WHO Expert Consultation on Foods Derived from Biotechnology was convened in 2000 to address food safety and nutritional questions regarding foods derived from GM plants, including a review of the scientific basis, application and limitations of the concept of substantial equivalence.

It concluded, based on the current application of substantial equivalence and alternative strategies that this concept contributed to a robust safety assessment framework. Moreover, the Consultation noted that substantial equivalence is a concept used to identify similarities and differences between GM food and a comparator with a history of safe food use which subsequently guides the safety assessment process.

While not a safety assessment per se, the substantial equivalence approach allows the structuring of the safety assessment through characterizing the similarities and identifying the differences which can then be the focus of further consideration. This approach has been referred to as a useful safety standard. It therefore permits inference that the new food under consideration will be no less safe than the conventional food under conditions of similar exposure, consumption patterns and processing practices. Substantial equivalence is therefore clearly not intended to be a measure of absolute safety, but instead recognizes that while demonstrating absolute safety is an impractical goal, demonstrating that there is reasonable assurance that the GMO-derived product under consideration is no less safe than a conventional food product is an achievable goal. Since it's development, the use of the substantial equivalence concept in the safety assessment of GMO-derived foods has been subject to misinterpretation and criticism.

The comparative nature of the substantial equivalence concept without prescribing the extent of the phenotypic and compositional comparisons has led some to criticize the concept as not being measurable, and therefore inappropriate in safety assessment. This and other criticisms relate, in part, to the mistaken perception that the determination of substantial equivalence was the end point of a safety assessment rather than the starting point.

After several OECD countries had gained experience with safety assessment of GMO-derived foods, an OECD workshop examined the effectiveness of the application of substantial equivalence in safety assessment. This workshop concluded that the substantial equivalence approach provides equal or increased assurance of the safety of foods derived from genetically modified plants, as compared with foods derived through conventional methods.

In 2000, the OECD Task Force for the Safety of Novel Foods and Feeds also reviewed the substantial equivalence concept, its interpretation and its application. While noting that the majority of international guidance documents addressing the safety assessment of genetically modified (GM) plants had interpreted substantial equivalence consistently, this Task Force reported that there were differences in how it was applied and that these differences needed to be resolved.

The Codex Ad Hoc Intergovernmental Task Force on Foods Derived from Biotechnology pursued the development of international guidance on the safety assessment of GMO-derived foods. To date, this Task Force has developed proposed draft principles and guidelines for the safety assessment of foods derived from modern biotechnology. These guidelines interpreted the concept of substantial equivalence as a way of structuring the safety assessment, consistent with the FAO/WHO interpretation.

The 'Draft Principles for the Risk Analysis of Foods Derived from Modern Biotechnology' and 'Draft Guideline for the Conduct of Food Safety Assessment of Foods Derived from Recombinant-DNA Plants' have been forwarded by the Codex Task Force to the 25th Session of the Codex Alimentarius Commission for adoption at Step 8 of the Codex procedure. A proposed draft guideline for the food safety assessment of foods produced using recombinant-DNA organisms continues to be developed. Substantial equivalence: the application

A conclusion from a 1995 WHO workshop on 'Application of the Principles of Substantial Equivalence to the Safety Evaluation of Foods or Food Components from Plants Derived by Modern Biotechnology' summarized the relationship between substantial equivalence and safety assessment as follows: Establishment of substantial equivalence is not a traditional safety assessment in itself, but a dynamic, analytical exercise in the assessment of the relative safety of a new food or food component to an existing food or food component. The application of the substantial equivalence concept is a key step in structuring the safety assessment of GMO-derived foods. Through appropriate analysis, the new food is compared phenotypically and compositionally to its conventional counterpart with a history of safe use. In this compositional comparison, the characteristics, including levels of key nutrients and toxicants, are considered relative to those of the conventional counterpart taking into account the natural variation for such characteristics.

When a genetic modification results in the insertion of a specific trait, the comparative analysis of the new food will identify both the intended effect (inserted trait) and the potential unintended effects of the modification. Further assessment can then focus on new or altered characteristics for which no history of safe use can be established.

One of the important benefits of applying the substantial equivalence concept is that it provides flexibility which can be a powerful tool in terms of food safety assessment. The comparative approach to structuring the safety assessment can be applied at several potential levels along the food continuum (i.e. harvested primary food material or unprocessed food product, individual processed fractions, or final food product or ingredient).

While from a practical point of view, the compositional comparison should typically be applied at the level of the unprocessed food product, the flexibility of the concept permits the determination to be targeted to the most appropriate level based upon the nature of the product under consideration. Where multiple fractions from a single source are destined to different food products, the comparative approach might be targeted at the level of the unprocessed food product in order to permit the safety assessment to apply to all derived fractions (e.g. for soybean, where multiple fractions are used widely in foods, assessment at the level of the seed is appropriate).

However, where a single fraction of a particular raw material is used as human food, then the comparison can focus at the level of the single fraction, thereby simplifying the safety assessment process (e.g. for canola, where the processed oil is the only fraction consumed by humans, safety assessment may appropriately be focused on comparison of the oil composition of the novel variety with traditional canola oil composition).

Application of the substantial equivalence concept in the safety assessment of a GMO-derived food depends on the identification of an appropriate comparator with an acceptable history of safe food use. It also requires that sufficient analytical data be available in the literature or be generated through analysis to permit an effective comparison. These requirements present a key limitation of the substantial equivalence concept since the consideration of similarities can only provide assurance of safety relative to those components assessed for the particular comparator. The choice of the comparator is therefore crucial to the effective application of substantial equivalence in establishing the safety of a GMO-derived food.

An appropriate comparator must have a well-documented history of use. If adverse effects have been associated with the particular food type, specific components of the food which are considered to be causative of those adverse effects should be described and well characterized in order to permit effective comparison. A joint FAO/WHO Expert Consultation on Biotechnology and Food Safety considered the application of substantial equivalence in the safety assessment of GMO-derived foods.

The consultation recommended that applicacation of the substantial equivalence concept entail consideration of the molecular characterization of the new food source; phenotypic characterization of the new food source in comparison to an appropriate comparator already in the food supply; and the compositional analysis of the new food source or the specific food product in comparison to the selected comparator. The guidance further elabourates on compositional comparison by highlighting what information should provide sufficient information to permit effective comparison.

The recommended focus of the compositional comparison is on analytical comparison of those components identified in the food source in question which are nutrients which provide a substantial impact in the overall diets (key nutrients) and toxicologically significant compounds known to be inherently present in the species (key toxicants). In addition, there is recognition that additional components might be identified for analysis based upon the molecular and phenotypic characterization and the nature of the genetic modification. In addition to key nutrients and toxicants, additional parameters may be appropriate for assessing the potential for unintended effects of a genetic modification.

When modifications are directed at metabolic pathways of key macro or micro nutrients, the possibility of an impact on nutritional value is increased, particularly if that food is a major dietary source of the nutrient affected. The potential for unintended effects would be determined, in part, by the nature of the intended alteration (e.g. consideration of fatty acid profile if an enzyme involved in fatty acid metabolism is introduced) and the data from molecular and phenotypic characterization. The consideration of key nutrients and key toxicants in the comparison which is essential to applying a substantial equivalence in safety assessments also introduces another limitation of the concept. The nature of the comparative approach with respect to nutrients limits its universality since the relevance of nutrients in a particular crop are dependent on consumption patterns which might vary from region to region. Where differences in consumption exist for a particular crop, these must be considered in the identification of the key nutrients for assessment. This is particularly true for crops which form a significant portion in the diet in a particular region. The comparative approach to assessment can be applied in each region, but conclusions for one region will not automatically hold for another region if there are significant differences in consumption patterns and processing practices. Another potential regional limitation is related to the application of the concept of substantial equivalence as opposed to an inherent limitation of the concept per se.

It may be difficult, particularly in developing countries, to apply the concept to assess the safety of foods where adequate nutritional databases are not available for a given population. A safety assessment using the substantial equivalence approach does not demonstrate that a GMO-derived product is identical to its conventional comparator since the compositional comparison does not take into account all components.

However, application of the guidance provides assurance that the comparison has considered those components most likely to be relevant to the safety of the product as it is expected to be consumed in a particular region. Recognizing the importance of there being compositional data available for applying substantial equivalence, the OECD Task Force for the Safety Assessment of Novel Foods and Feeds has focused on the development of science-based consensus documents containing information on the nutrients, anti-nutrients or toxicants, product use and other data relevant to the assessment.

The OECD have published such documents for potatoes, sugar beet, soybean and low erucic acid rapeseed (canola). Such guidance will add to the understanding of appropriate parameters for a comparative assessment of composition while recognizing that additional parameters may be relevant to the safety assessment dependent on differences in consumption pattern. The international guidance developed by the OECD and FAO/WHO has been practically applied to the safety assessment of GMO-derived food products in several countries. In order to facilitate such assessments, specific guidance documents which embrace the substantial equivalence concept have been published. The majorities of these currently address the safety assessment of genetically modified plants and have consistently interpreted the concept of substantial equivalence. Internationally, these guidance documents have been applied to the assessment of a significant number of GMO-derived plant products over a period of more than eight years, demonstrating that the concept of substantial equivalence can be applied effectively in the safety assessment of novel foods.

GENETICALLY MODIFIED FOODS

Increasingly, genetically modified foods are making their way to the marketplace. Since they are not labeled, consumers generally do not know that they are eating such products. Not everyone is comfortable with the present system. A few years ago, one rarely heard about genetically modified organisms (GMOs). In fact, few people knew exactly what they were. That is no longer the case. GMOs, or plants, animals, or microorganisms that have been genetically altered or engineered for a specific purpose, are everywhere.

It has been estimated that almost half of all American farmers grow some variety of genetically modified crops, also known as genetically engineered crops. The Food and Drug Administration (FDA) says that GMOs are filling the shelves of stores. "Tomatoes, potatoes, squash, corn, and soybeans have been genetically altered through the emerging science of biotechnology.

So have ingredients in everything from ketchup and cola to hamburger buns and cake mixes". In Genetically Engineered Food, Martin Teitel, executive director of the Council for Responsible Genetics, and Kimberly A. Wilson, director of the council's programme on Commercial Biotechnology and the Environment, describe the sweeping changes brought about by GMOs.

"The genetic engineering of our food is the most radical transformation in our diet since the invention of agriculture 10,000 years ago. These food crops are already growing on millions of acres all around the world".

Similar sentiments were expressed in Genetically Engineered Food by Ronnie Cummins, national director of the Organic Consumers Association, and Ben Lilliston, a health and environment writer and communications coordinator for the Institute for Agriculture and Trade Policy in Minneapolis: "Genetic engineering...is a revolutionary new technology still in the early experimental stages of development.

It enables molecular biologists to permanently alter the essential characteristics or genetic codes of living organisms. This technology has the awesome power to break down fundamental genetic barriers-not only between species but between humans, animals, and plants. For the first time in history, the scientists and corporations using this technology have become, in effect, the architects, builders, and 'owners' of life".

Introducing the genes of one type of food into another, thereby changing the genetic structure of an organism, gives the second plant or animal beneficial characteristics, such as resistance to disease, improved nutritional value, and better growth. These modifications enable foods to grow faster, stronger, or bigger, all while using far less pesticide. There are now tomatoes that last longer and corn that is resistant to pests.

"The acreage devoted to herbicide-resistant crops has been expanding because planting them reduces the need to plow more ground, decreases the amount of herbicidal chemicals needed, produces higher yields, and can deliver a higher grade of grain and other products". Transgenic or genetically modified crops "can be developed for pest resistance, improved yield, tolerance to biotic and abiotic stresses, use of marginalized land, improved nutritional content, and the production of vaccines and pharmaceuticals".

A 2000 story in Popular Science said, "Genes from flounders can help ordinary plants like tomatoes and strawberries fight the cold. Researchers are also inserting bacterial genes into corn and soybean plants to better protect them from insects or render them immune to certain herbicides". Cross-breeding has been around for a long time. "Virtually all common fruits and vegetables look and taste the way they do because of hybridization".

However, biotechnology has added a whole new dimension to selective breeding. "This process allows for the transfer of only one or a few desirable genes, thereby permitting scientists to develop crops with specific beneficial traits and those without undesirable traits.

Current technology permits scientists to alter one plant characteristic at a time, thereby not spending years trying to develop the best tasting and hardiest plants". For example, Craig Nessler, professor and head of the Department of Plant Pathology, Physiology, and Weed Science at Virginia Tech University, inserted genes from rats into lettuce seeds. This produced leafy greens with much higher amounts of vitamin C.

"Rats, it turns out, carry a gene that allows their bodies to manufacture vitamin C, which is why the little four-legged fiends never develop scurvy on long sea voyages". Years ago, such science would have been unthinkable. The differences between traditional breeding and genetic engineering. "To give a simple example, a traditional breeder interested in producing a yellow tomato must find the yellow trait in a plant that will breed with the tomato by natural mechanisms.

The only plants that can breed with tomatoes are closely related ones. Unrelated plants like oak trees or cantaloupes could not breed with tomatoes, and thus could not contribute new genes. A genetic engineer, on the other hand, can consider any organism-even a butterfly or a daffodil-as a source of the yellow trait. If the gene that determines yellow colour has been identified and isolated, it can be directly transferred into tomato plants".

The International Food Information Council maintains that because GMOs reduce the amounts of pesticide used by farmers, they are healthier for the environment. "Insect protected crops allow for less potential exposure of farmers and groundwater to chemical residues, while providing farmers with season-long control. Also by reducing the need for pest control, time, effort and resources spent on the land are less, thereby preserving the topsoil". It is almost impossible for the average consumer to know whether there are GMOs in his or her food.

Playing on consumers' fear of the unknown, a few innovative marketers include "No GMOs" notations on their labels. However, the vast majority of products do not indicate whether they are comprised of any genetically altered ingredients. "Legal authority for food labeling rests with the Food and Drug Administration. Foods derived from biotechnology currently must be labeled only if they differ significantly from their conventional counterparts-for example, if their nutritional content or potential to cause allergic reactions is altered".

SUPPORTERS AND OPPONENTS OF GMOS

Should consumers feel comfortable eating genetically altered food? The controversy surrounding GMOs, or Franken foods, as they are called by those who oppose them, is filled with heated discussion. In general, the federal government, big business, and many researchers and numerous well-respected publications maintain that genetically altered food is perfectly safe. They contend that GMOs offer the potential to feed healthier food to far larger numbers of people.

That is of particular significance, they say, when considering the vast amounts of food needed to feed the populations of developing countries. About 15% of the world's population-about 800 million people-consume less than 2,000 calories per day. They are always hungry and live in a state of chronic malnourishment. Large numbers of these people are women and children. "More than 180 million children under five years of age are underweight, that is, they are more than two standard deviations below the standard weight for their age. This represents one-third of the under-fives in the developing countries."

These groups are growing at a staggering rate. "By the year 2020, there will be about an extra 1.5 billion mouths to feed. If the proportion of the population of the developing countries deprived of an adequate diet remains the same, the number undernourished 20 years from now could be well over one billion". Not everyone agrees. Sizeable numbers of equally esteemed researchers and consumer and environmental groups worry about a variety of factors. "Concerns include ethical issues related to potential long-term health effects of eating bioengineered foods, labeling, and potential environmental risks".

In Europe, particularly Great Britain, large numbers of people have supported a ban of GMOs. Grocery stores are refusing to carry foods containing GMOs. Other countries are following Britain's lead. Moreover, concern about risk of lawsuits and decline in land value has made rural land managers hesitant to authorize tenants to grow GM crops." Within the European Community, there is enormous apprehension.

"Consumers worry about the impact of any health risks that might eventually be discovered, especially if whole national populations have ingested GM food. Health effects, moreover, may stem from risk related to questions that scientists have not yet asked rather than those questions for which reassuring answers have been found".

In High-Tech Harvest, Elizabeth L. Marshall, a science writer, notes that the most serious GMO controversy has centered on food labeling. Usually, opponents of GMOs favour labeling. They want to know which foods have been changed and believe that shoppers, when presented with such information on the label, would shy away from them. "But some supporters want labels too. They believe that labels would help pave the way to consumer acceptance". Genetic engineering food research began in the 1980s. In 1990, the enzyme chymosin was the first genetically engineered food to receive the approval of the FDA. More than half of the hard cheese currently sold in the United States is made with chymosin produced by genetically engineered fungi. The rest is made with chymosin obtained from rennet, an enzyme from the stomachs of slaughtered calves.

"Although cheese-makers can use chymosin from either source, the chymosin made from genetically engineered bacteria is easier to obtain and more pure". Two years later, in 1992, the FDA approved the first genetically engineered whole food-the Calgene Flavr Savr tomato, which became available for sale in 1994. Because of a change to a single gene, the Calgene Flavr Savr tomato ripened without getting soft. However, the Flavr Savr tomato encountered a number of problems, such as an inability to tolerate shipping.

"Contrary to Calgene's expectations, the tomatoes were often so soft and bruised that they could not be sold as fresh produce, and most of the varieties did not have acceptable yields or disease resistance in tomato-growing regions". The Flavr Savr marketing campaign was unsuccessful. The tomatoes received a good deal of negative press, and consumers refused to buy them. "Calgene even tried marketing the tomatoes as a gourmet product under the friendly sounding 'MacGregor's' brand name, but despite their presence in thousands of U.S. grocery stores, consumers did not want to pay more for genetically engineered tomatoes."

In 1996, the Flavr Savr tomato was removed from the market. Teitel and Wilson noted that "the Flavr Savr grew well in the laboratory but encountered serious obstacles in the field". Nevertheless, genetic engineering moved forward. By 1996, genetically engineered seeds for corn and soybeans were available, followed by seeds for potatoes, squash, tomatoes, and rapeseed. "Each type of seed offers a characteristic that natural seeds did not. Some protect the plants against insects or disease, while others delay ripening or protect the plants from chemicals used to kill weeds.

REGULATION

It is important to realise that genetic engineering is a tool that may be used in a variety of ways to make many different food products. That is why, Marshall contends, the FDA decided in 1992 that genetically engineered foods would be reviewed "based on their individual safety and nutrition, rather than on the methods used to produce them." The FDA deemed them "as safe as foods developed through other agricultural technologies".

As a result, the FDA regulates GMOs as it does any other food. If a new product contains food items that are already considered safe, the manufacturer does not need to obtain special permission to sell the item. "In other words, a transgenic [another word for genetically engineered] catfish containing a trout gene would be considered safe to eat because it is already well-known that trout is harmless to humans. The only time the law requires testing of a new food is when it contains an additive that is not known and, therefore, not generally regarded as safe". In 1992, the FDA determined that there would be exceptions to this rule.

If a new food contained ingredients that were likely to trigger allergic reactions in allergic individuals or if the manufacturing process dramatically altered the nutritional content, a label would be required. Unfortunately, this is not as clear-cut as it may initially appear to be. Foods are constantly combined together to form new foods. Should everything be labeled? The government and sizeable numbers of people within the food industry say that such labeling would be confusing and expensive.

In Pandora's Picnic Basket, Alan McHughen, a senior research scientist at the University of Saskatchewan, Canada, wrote that "in North America, regulators and companies agree that mandatory labels should be reserved for those products carrying a documented health risk or substantial change in nutritional composition. If the GM products are 'substantially equivalent' to conventional counterparts, the companies argue, the GM label would be misleading". Nonetheless, many consumer groups consider labeling to be an absolute necessity.

EarthSave International explains that genetic engineering "introduces new proteins into the human and animal food chains. This means that human beings are now consuming products that have never before been considered foodstuffs. There is concern that these new proteins could potentially cause toxic or allergic reactions or other health effects. Unfortunately, there is no easy way to predict the allergenic potential... since allergic reactions typically occur only after the individual consuming the food is sensitized by initial exposure to the allergen".

It is true that the government has mandated the labeling of other products such as foods that are processed and those containing sulfites. Labels must note the source of hydrolyzed proteins, and they are required to state if a food-other than a spice-has been irradiated. Cigarette labels have a warning from the surgeon general. "These legal requirements are in place because large numbers of citizens want such information, and a specific fraction needs it.

An identifiable fraction of consumers actually needs information about genetic modifications-for example, regarding allergenicity and almost all consumers want it". Lawrence Kushi, an epidemiologist at Columbia University, notes that it is simply untrue to say that genetically modified food is no different from the conventional equivalents. He maintains that labeling such foods would give consumers the information they want while allowing scientists to examine whether they have any negative impacts on human health or the environment.

FAVOURable Condition for GMOS

Some people contend that the negative publicity that consumer groups generate about GMOs is most harmful to the world's most vulnerable people-the poor. They say that unlike residents of wealthier countries, large segments of the population of developing countries worry about dying from malnutrition and starvation. By using GMOs, farmers in developing countries may harvest more crops with less pesticide.

Playing on this theme, frequently appearing television and magazine ads have featured rice that has been genetically altered with a daffodil gene. As a result of that modification, the rice contains beta-carotene, which the body converts into vitamin A. The ads claim that by eating this form of "Golden Rice," countless numbers of people in poorer countries will be spared blindness. About 400 million people may be in danger of vitamin A deficiency. Of these, 100 to 200 million are children. Mammals do not manufacture vitamin A, so it must be obtained through the diet.

People who eat lots of fruits and vegetables and animal products obtain adequate amounts. Millions of people are not as fortunate. "1.0 to 2.5 million deaths per year of preschool children-up to 30% of total deaths in that age group-could potentially be averted by bringing vitamin A deficiency under control worldwide. Combined with expanded vitamin A supplementation programs-which will continue to be important-Golden Rice is expected to make a major contribution to improving the health of millions of the world's poorest children". For the unspoken challenge here is that if we don't get over our queasiness about eating genetically modified food, kids in the third world will go blind." But Pollan asserts that there is far more to the Golden Rice issue than is initially evident. For example, in order to meet the minimum daily requirement for vitamin A, it would be necessary for an average 11-year-old to eat 15 pounds of cooked Golden Rice. Moreover, in order to convert beta-carotene into vitamin A, fat and protein must be present, and these are lacking in the diet of malnourished children. Pollan also wonders whether Asians will accept rice that is golden in colour.

It is already well known that brown rice is healthier than white rice, yet Asians consistently prefer white rice. Apparently, at some point, Confucius "extolled the pure whiteness of rice as the ideal backdrop for green vegetables". Pollan further questions the ethics of an industry that is using the suffering of children to sell its food. In "Vitamin A Deficiency Disorders: Origins of the Problem and Approaches to Its Control," Alfred Sommer writes that there are some "hurdles" to be overcome before Golden Rice may have an impact on society. "The strains must be able to grow under the varied conditions in countries with vitamin A deficient populations. The yield and the cost must be attractive to the farmer (or benefit from public sector subsidization). The organoleptic [sensory] qualities of the rice must be acceptable to the target population (women and children). The beta-carotene needs to be bioavailable, the degree dependent on its concentration in the rice, the matrices to which it is bound, the effect of traditional cooking methods and the amount consumed".

Sommer warns that Golden Rice will never be a complete solution to the problem of vitamin A deficiency. "Many deficient populations do not consume rice, and even within traditional rice-consuming countries, some high-risk groups will not be able to afford it". Regardless, it is impossible to deny that the population of the world is growing at a sometimes mind-boggling rate. Presently, the world has about 5 billion people. By the year 2050, that figure will double to 10 billion. Large numbers of these people will live in the developing world, where hunger is rampant.

The International Food Information Council states that "by increasing a crop's ability to withstand environmental factors, growers will be able to farm in parts of the world currently unsuitable for crop production. Along with additional food, this could provide the economies of developing nations with much-needed jobs and greater productivity". Feeding so many people takes a staggering toll on the environment. "Erosion can claim precious topsoil, farm chemicals sometimes reach streams, rivers and ground water supplies, and livestock can deplete grazing lands.

Wetlands and other sensitive habitats sometimes get plowed under for use as farmlands. And, in the world's tropical forests where an estimated 90 per cent of the world's species exist, poor farmers clear trees in order to provide food and a living for their families." The International Food Information Council believes that GMOs may play a pivotal role in mitigating this burden. Crops that have been genetically modified are often more resistant to disease and require a reduced level of insect control.

Therefore, fewer acres need to be planted to reap the same amount of food. One of the key players in the effort to increase the production of rice in Asia has been the Rockefeller Foundation. The foundation believes that GMOs will enable Asian farmers to obtain more rice from the same amount of land. "By the year 2005, the Foundation's hope is to increase rice production in Asia by 20 per cent through the use of biotechnology without degrading the environment or reducing farm incomes."

The Rockefeller Foundation has also been conducting similar work in Africa. Former President Jimmy Carter, who has become well known for his humanitarian works, has observed that "responsible biotechnology is not the enemy; starvation is. Without adequate food supplies at affordable prices, we cannot expect world health or peace". Earth Save International disagrees with the notion that genetic engineering may play an important role in feeding the world's growing population. "World hunger is not a problem of technology or insufficient production, but primarily one of unequal distribution and economic inequality. As farmers lose their land and move to the cities, they also lose their food-independence and begin to rely on money, often in dramatically short supply for many in the third world, in order to buy food that they formerly grew themselves". In The Ecological Risks of Engineered Crops, Jane Rissler and Margaret Mellon agree.

"Even if research is done on the right crops and result in increased agricultural productivity, increased production is only one factor in the complicated equation of world hunger. Poverty, trade policies, subsidies, soil erosion, and water shortages are also important causes of hunger.

Increases in productivity obtained through genetic engineering will have mixed effects if not developed with due regard to the other important aspects of the hunger problem". In The Future of Food, Brian J. Ford, an English scientist, writer, and television host, says that a number of everyday farm animals and crops-such as pigs, cattle, wheat, oats, barley, and rye-are the result of human intervention. "Traditional farmers have been producing new animals and plants by cross-breeding for ten thousand years."

While genetic engineering "offers a more radical way of manipulating characteristics," it is fast becoming a part of daily life. "Genetic modification is inevitable. Like electric power, road transport, or computers, it is a facet of the future and the public will gain little by campaigning to ban this potentially rewarding technology. Properly applied, it could offer us so much".

All the same, Ford does not believe that the public should simply accept whatever the bioengineering companies do. Rather, he calls for a careful monitoring of the industry and the introduction of safeguards. "This is a huge new industry, and it will have pronounced effects on us all. We will need to control it". Ford suggests the following controls:

- Approval for new experiments should always be sought from a regulatory authority well versed in the subject.
- None of the members of such a committee should be in a position to benefit commercially from approval.
- No genes conferring problematical properties (e.g., antibiotic resistance) shall be used, whether as markers or otherwise, outside enclosed laboratories.
- Agents capable of transfer to wild plants, like pollen-carrying genes conferring resistance, shall not be liberated into the environment.
- The public shall be consulted about the siting and the benefits of experiments.
- All GMO products should be properly labeled and the source declared.
- Records of possible unwanted side effects should be meticulously maintained. All such events should be investigated by an organization unconnected with the source of the product.

ARGUMENTS AGAINST GMOS

The controversy shows no sign of diminishing. GMO opponents present some compelling arguments. The Organic Consumers Association contends that genetic engineering deconstructs basic, fundamental genetic barriers between species. "By combining the genes of dissimilar and unrelated species, permanently altering their genetic codes, novel organisms are created that will pass the genetic changes onto their offspring through heredity.

Animal genes and even human genes are being inserted into plants or animals creating unimagined transgenic life". Completely new organisms will be created. The association believes that these have the potential to compromise human health and well-being, animal welfare, and the environment. The following are a few examples of ways in which this might occur:

- Genetically engineered organisms can reproduce, migrate, and mutate. If they escape or are released from the laboratory, they could wreak environmental havoc.
- Gene splicing can have unpredictable and dangerous results.
- Genetic engineering of crops and food-producing animals can produce toxic and allergic reactions in humans.
- Safety testing and regulation of genetically engineered organisms are inadequate or lacking.
- Patenting genes they discover and living organisms they create will enable a small corporate elite to own and control the genetic heritage of the planet.

The poll found that only about one-third of the Americans surveyed view GMOs as safe to eat. A little more than half found them unsafe for human consumption. Women, who tend to do the family shopping, were even more against GMOs than men. While 40% of men think that GMOs are unsafe, 62% of women feel that way. A striking percentage of those surveyed-93%-want the federal government to require food labels to indicate the presence of GMOs.

Fifty-seven per cent said that they were less likely to purchase food that contains GMOs. There are even age differences. "People under 45 are about 10 points more likely than their elders to think genetically modified foods are safe to eat. But a majority of young adults still calls genetically modified foods unsafe". In Brave New Seeds, Robert Ali Brac de la Perrière, an international consultant, and Franck Seuret, a journalist, contend that throughout the world, GMOs will weaken the role of small-scale farmers who lack the wherewithal to pay higher prices for genetically modified seeds and lack the expertise that more technically oriented farming demands. Increasingly, large farms will absorb the smaller ones. Further, farmers are losing control over their products. "With transgenic plants, the freedom of agricultural practice and liberty of choice is reduced as the farmer works with a patented product which is subject to very specific conditions of use.

For example, contracts for the use of a transgenic soya variety that is resistant to the herbicide RoundUp forbids, amongst other things, the cultivation of other varieties, the use of herbicides other than those allowed by Monsanto and the exchange of seeds with neighbours". Crop diversification reduces the amount of herbicide and synthetic nitrogen fertilizer that may be needed, improves the quality of soil and water, controls insect-pest and pathogen populations, increases crop yield, and reduces yield variance.

Insect-resistant crops may have unintended victims and disrupt other forms of pest control. Also, toxins from these products may remain in the soil for extended periods of time and reduce the level of soil fertility. As if this were not sufficiently problematic, there is the issue of "terminator" seeds, genetically altered plants that are programmed to kill their own seeds. Unlike most plants, these do not regenerate. "After the seeds are planted and the crop matures, the [seed-killing] toxin is produced, killing the new seeds the plants carry".

In the previously noted Genetically Engineered Food, Martin Teitel and Kimberly A. Wilson say that terminator seeds end the long-held practice of saving seeds for future plantings. "Farmers who use terminator seeds will be forced to buy new seeds for every planting, ending an age-old tradition of seed saving and creating a perpetual cycle of dependence on big seed companies". In Brave New Seeds, Brac de la Perrière and Seuret observe that terminator seeds "achieved their supreme goal-enslaving their clients".

There is an even more alarming possibility. "Some doomsday scenarios suggest pollen from Terminator plants could drift with the wind like a toxic cloud, cross with ordinary crops or wild plants, and spread from species to species until flora all around the world had been suddenly and irreversibly sterilized". That is probably a huge exaggeration. Gene drift does not occur that easily. A large number of crops pollinate themselves. As for those that do not, border fields may be established to contain genes. In "Genetically Modified Foods: Are They a Risk to Human/Animal Health?," Generally, when testing products, the industry has compared a GM food to a non-GM food.

"When they are not significantly different the two are regarded as 'substantially equivalent,' and therefore the GM food crop is regarded as safe as its conventional counterpart." Pusztai said that "substantial equivalence is an unscientific concept that has never been properly defined and there are no legally binding rules on how to establish it."

Additionally, there are currently no reliable methods to test new genetically engineered crops for food allergies. "It is at present impossible to definitely establish whether a new GM crop is allergenic or not before its release into the human/animal food/feed chain".

EVALUATION OF GENETICALLY MODIFIED FOOD CROPS

Genetic modification, otherwise referred to as recombinant DNA (rDNA) technology or gene-splicing, has proven to be a more precise, predictable and better understood method for the manipulation of genetic material than previously attained through conventional plant breeding. To date, agricultural applications of the technology have involved the insertion of genes for desirable agronomic traits (*e.g.* herbicide tolerance, insect resistance) into a variety of crop plants, and from a variety of biological sources. Examples include soybeans modified with gene sequences from a *Streptomyces* species encoding enzymes that confer herbicide tolerance, and corn plants modified to express the insecticidal protein of an indigenous soil microorganism, *Bacillus thuringiensis* (Bt). A growing body of evidence suggests that the technology may be used to make enhancements to not only the agronomic properties, but the food, nutritional, industrial and medicinal attributes of genetically modified (GM) crops.

Regulatory supervision of rDNA technology and its products has been in place for a longer period of time in the United States than in most other parts of the world. The methods and approaches established to evaluate the safety of products developed using rDNA technology continue to evolve in response to the increasing availability of new scientific information. As our understanding of the potential applications of the technology is broadened, the safety of products developed using rDNA technology and the potential effects of introduced gene sequences on human health or the environment will be more closely scrutinized. In fact, much of the knowledge acquired during the commercialization of the products of rDNA technology in agriculture is now finding application in evaluating the safety of products developed through more conventional means.

The objective of this chapter is to provide the reader with an overview of the significant events leading up to the present, science-based, regulatory framework that exists for the safety evaluation of GM food crops within the United States. An attempt has been made to discuss concerns over the sufficiency of existing regulations, as well as to highlight recent initiatives taken by federal regulatory agencies to address them. Through better communication of how the regulatory process functions within the United States, it is anticipated that current and future applications of rDNA technology in agriculture will be met with a greater level of understanding and acceptance.

HISTORICAL PERSPECTIVE

rDNA technology was first developed in the 1970s. The initial response of the scientific community, including members of the National Academy of Science (NAS), to the prospects of rDNA technology, was to postpone any further research involving the technology until the potential risks to human health and the environment could be evaluated. Researchers attending the International Conference on Recombinant DNA Molecules in 1975, otherwise known as the Asilomar Conference, tried to establish a scientific consensus on how best to self-regulate emerging applications of the technology. The conditions and restrictions that were proposed at this conference have formed the basis by which federal guidelines and policies for rDNA technology research were drafted within the United States.

National Institutes of Health (NIH)

The National Institutes of Health (NIH) was the first federal regulatory agency to publish their interests in evaluating the safety of rDNA technology in 1976, in the form of guidelines for the conduct of research. Because of the uncertainties that existed at the time, all research into the potential applications of rDNA technology was limited to the confines of federally funded labouratories under NIH control. After continued research, and a more careful assessment and monitoring of the risks, a set of less restrictive guidelines was published in 1978.

However, the environmental release of organisms developed using rDNA technology outside the confines of controlled labouratory conditions was prohibited unless otherwise approved by the NIH director. In the early 1980s, the NIH established an rDNA Advisory Committee (RAC) to review all data and experience gained with applications of the technology under its control. Based on recommendations of the RAC, a more relaxed set of research guidelines was published by the NIH in 1983.

The NIH approved the first environmental release of an organism developed using rDNA technology (ice-minus strain of *Pseudomonas*) in 1983. In response, they were criticized for failing to prepare a statement or assessment of the environmental impact of their regulatory decision as required under the National Environmental Policy Act (NEPA). Once the legal controversy had subsided, all responsibility that the NIH had for regulating the environmental introduction of GM organisms was relinquished. Nevertheless, NIH guidelines continue to be referenced in assessing the safety of rDNA research performed within industry, federal and other state labouratories. However, it was unclear which federal regulatory agencies would be responsible for ensuring the safety of the products developed using rDNA technology.

Office of Science and Technology Policy (OSTP)

In response to a need for clarification, the Office of Science and Technology Policy (OSTP) began work on the development of a policy to establish a federal regulatory framework for evaluating the safety of products developed using rDNA technology. Following an opportunity for public comment, OSTP published a final version of the 'Coordinated Framework for Regulation of Biotechnology' (Coordinated Framework) in 1986. The policy provided the basis by which federal regulatory agencies got involved in evaluating the safety of products at later stages of commercial development at that time.

Table. Overview of Responsible Agencies under the Coordinated Framework

Responsible Agency	Products regulated	Reviews for safety
FDA	Food, feed, food additives, veterinary drugs	Safe to eat
USDA	Plant pests, plants, veterinary biologic	Safe to grow
EPA	Microbial/plant pesticides, new uses of existing pesticides, novel microorganisms Safe for the environment	Safety of a new use of a companion herbicide

According to the Coordinated Framework, the products of rDNA technology should be regulated on the basis of the unique characteristics and features that they exhibit, not their method of production. The products of rDNA technology were considered to pose risks to human health and the environment similar to those posed by conventional products already regulated within the United States. As a result, no new federal regulatory agencies or regulations were required. The Coordinated Framework did not, however, rule out the possibility of the development of new guidelines, procedures, criteria or even regulations to supplement or alter the scope of existing statutes for the products of rDNA technology. The Coordinated Framework identified three federal regulatory agencies within the United States: the US Food and Drug Administration (US FDA), the US Department of Agriculture (USDA) and the US Environmental Protection Agency (US EPA), as having primary responsibilities for evaluating the products of rDNA technology under development at that time.

In 1992, the OSTP released another document entitled, 'Exercise of Federal Oversight within the Scope of Statutory Authority: Planned Introductions of Biotechnology Products Into the Environment', outlining the proper basis by which federal regulatory agencies were expected to exercise their regulatory authority. As with conventional products, dependent upon the intended use and function, more than one federal regulatory agency may share an interest in evaluating the safety of a product developed using rDNA technology. If more than one federal regulatory agency has an interest, lead agencies are identified as being responsible for coordinating activities to limit any potential duplication of efforts. Although federal regulatory agencies worked independent of one another, it was realised that close working relationships would need to be established in order to evaluate effectively the safety of products developed using rDNA technology. Recently, the OSTP teamed up with the White House Council on Environmental Quality (CEQ) to perform a six-month inter-agency evaluation of the federal regulatory agency responsibilities in evaluating the environmental safety of products developed using rDNA technology. A case-study approach for a variety of different classes of products developed using rDNA technology was used to evaluate the level of federal regulatory agency involvement, to identify strengths, weaknesses and areas of potential improvement. The review concluded that none of the previously approved products of rDNA technology has had any significant negative impact on the environment.

Although all the case studies were published, OSTP/CEQ failed to reach a consensus on issues relating to the relevant strengths and weaknesses of the existing regulatory structure within the time allotted for the completion of its review. A review of the case studies published provides a comprehensive interpretation of the responsibilities of each federal regulatory agency in ensuring the safety of the products developed using rDNA technology.

National Academy of Sciences (NAS)

The National Academy of Sciences (NAS), and its operating arm, the National Research Council (NRC), have served as a primary source of scientific, technological, human health and environmental policy advice during the development of regulatory approaches for the safety evaluation of products developed using rDNA technology within the United States.

In 1987, the NRC published a report concerning the potential human health and environmental hazards associated with the commercial introduction of GM organisms, entitled *Introduction of Recombinant DNA-Engineered Organisms into the Environment: Key Issues*. The risks associated with the introduction of GM organisms were considered to be essentially the same in kind as those associated with unmodified organisms.

In other words, rDNA technology did not appear to introduce any unique risks as compared to the products that had been developed using more conventional methods of genetic modification. In reaching these conclusions, the NRC performed an evaluation of the similarities and differences in the properties exhibited by products developed using a variety of different techniques. To this day, the conclusions of this report continue to be referenced by the regulators and developers of GM crop varieties worldwide.

A subsequent NRC report, entitled *Field Testing Genetically Modified Organisms: Framework for Decisions*, also reached similar conclusions, but provided additional guidance as to how regulatory decisions concerning the introduction of GM organisms should be made. The NRC recommended that regulatory decisions concerning the introduction of GM organisms should be made on a case-by-case basis. Consistent with the Coordinated Framework, the NRC did not consider the nature of the process used for the genetic modification of an organism to be a useful criterion for determining whether a product requires less or more regulatory oversight. As a result, no valid reason existed to regulate organisms genetically modified via modern techniques (*e.g.* rDNA technology) any differently from organisms genetically modified via more conventional means. Similar conclusions have been published in the reports of international standards-setting organizations.

In retrospect, within both reports, the NRC acknowledged that modern and conventional methods of genetic modification are not without risks to human health or the environment, and as a result, neither could be considered inherently more risky. As a result, regulatory decisions concerning the safety of the products of rDNA technology need only to take into consideration the specific characteristics exhibited by a particular GM organism and the environment in which it is to be introduced, and not the method by which it has been produced. The NRC has further articulated their conclusions into what is now commonly referred to as the 'Concept of Familiarity'.

Although familiarity with the characteristics of a particular organism or the environment to which it will be introduced would not necessarily mean it was safe, it can be expected to provide a sufficient amount of information to allow for a judgement to be made of the risks. For example, familiarity with a new GM plant variety could be established based on comparisons between characteristics of the parent line or other crop species exhibiting similar traits, as well as through the results of actual field tests involving the GM plant. These principles were further elabourated upon by the Organisation for Economic Co-operation and Development (OECD), and as a result, have been referenced in the development of regulatory policies for evaluating the safety of GM crops on a global basis.

More recently, a committee established by the NRC published a report entitled *Genetically Modified Pest-Protected Plants: Science and Regulation,* based on a review of all scientific and regulatory data collected during the regulatory approval process for GM crops within the United States. The primary objective was to assess independently the effectiveness of existing and proposed regulations for the safety evaluation of GM crops expressing plant pesticides (*i.e.* plant-incorporated protectants). No new evidence was identified to suggest plants expressing plant pesticides posed any greater risk to human health or the environment as a result of their genetic modification. In fact, the NRC concluded, 'with careful planning and appropriate regulatory oversight, commercial cultivation of GM plants is not expected to pose higher risks and may pose less risk than other commonly used chemical and biological pest-management techniques'.

However, the NRC report included requests for federal regulatory agencies to further strengthen the current regulatory approval process through better coordination and communication between agencies, on-going investment in the research and monitoring of potential human health (*e.g.* allergenicity) and environmental impacts (*e.g.* insect resistance), and by providing greater access to information evaluated in support of regulatory decisions.

CONSUMERS SUPPORT GM FOOD

Many Asians are not opposed to the concept of biotechnology. People are enthusiastic about the potential for improved food quality and safety through biotechnology. Asian consumers remain positive about the many potential benefits of GM food, despite unbalanced media coverage and confusion in the international marketplace. A survey coordinated by the Asian Food Information Centre in Dec 1998- Nov 1999 revealed the following: More than 80% of people were prepared to try a food product made from genetically modified crops. More than half of the respondents would purchase or eat genetically modified foods if they were proven to be healthier.

Farmer Benefits

Biotechnology could offer farmers a more efficient way to produce safer products in greater numbers. Some of the benefits include: More flexibility in operations, lesshard manual labour. Efficient pest control built into crops means farmers are exposed to less risk and more willing to invest in their farm operations. Farmers can control weeds at any time. Finally, it makes farming more attractive to the younger generation.

Environmental Benefits

- Higher yields decrease need to convert forests and natural habitat into farmland.
- *Conservation tillage aided by biotechnology*: The development of herbicide resistant crops has allowed farmers to practice conservation-tillage farming. This is the practice of planting seeds through the stubble of last year's crop, rather than plowing and disking the field. The stubble protects topsoil against loss to wind and rain and reduces chemical run-off to streams. By not plowing, farmers also conserve soil moisture, which can reduce irrigation demands in some regions.
- *Improved water quality through reduced soil erosion and run-off*: Soil sedimentation or siltation is sometimes a major threat to stream quality. A summary of studies found that no till farming which is facilitated by biotechnology, can reduce soil erosion by 90%.
- Reduced use of insecticides and herbicides aids beneficial insects, wildlife, plants, and humans - Use of biotechnology derived plants with built-in resistance to insects and diseases could mean that substantially fewer pesticides, insecticides, and herbicides would be required, thus reducing farmers' exposure, a particular problem in many developing countries where protective gear may not always be available.

- *Nitrogen fixing crops could reduce use of nitrogen fertilizers*: One of the biggest challenges in agriculture is the fact that most crops, with the exception of legumes such as clover, do not naturally produce nitrogen, but require the application of nitrogen fertilizers for growth. Considered "frontier" research at present, international scientists' research is progressing to produce nitrogen-fixing rice. That is, rice that can produce its own nitrogen to enable it to grow. Currently, it is estimated that about 10 million tons of nitrogen fertilizer are needed each year for rice production worldwide, and that amount is expected to double in the next several decades. Besides the potential decreased cost to poor farmers, the massive amounts of fuel needed each year to produce nitrogen fertilizers could significantly decrease with the production of such rice. Scientists have made progress in transferring one gene involved in the nitrogen fixing process to rice plants, and are working on achieving the same results with other genes involved.
- *Biodiversity can be aided by improved plant resilience*: Biotechnology can aid in preserving endangered plants through genetic enhancement of traits needed for survival. Modern biotechnology, together with other methods, such as seed banks, can help to preserve and even reintroduce native plants that are endangered. This could promote biodiversity and help to enhance food security by saving plant varieties for the future.

Economic Benefits

In the US, the Primary Benefit of Bt Corn Varieties has been Increased Yields

In 1999, it is estimated that 66 million bushels of corn were saved from the corn borer, or the equivalent production of nearly 500,000 acres.

- Net revenues are estimated to have increased by U$99 million in 1999.
- The use of herbicide tolerant soybean varieties has been a reduction in weed control costs, of U$216 million per year in 1999 and growers have also reduced the number of herbicide applications, by 19 million in 1999.

In addition to the potential consumer benefits, agricultural biotechnology presents a range of possible economic benefits.

Impact of Bt Cotton in China

- A sample of 283 cotton farmers in Northern China was surveyed in December 1999. Farmers that used Bt cotton reduced the use of pesticide without reducing output/ha or quality of cotton. This resulted in substantial economic benefits for small farmers.

- At least 85% of the 1999 benefits from the adoption of Chinese Bt cotton varieties went to small scale farmers
- Farmers who grew most popular Bt varieties reduced their costs of production by 20-23% over new non-Bt varieties.
- Small farmers- those whose farms are less than 1 ha or have family incomes less that RMB 10,000-gained almost twice as much income per unit of land from adopting Bt cotton as large, more wealthy farmers gained.
- In addition to the potential consumer, environmental and economic benefits of biotechnology, there should be worldwide benefits in meeting the food needs of the growing population in Asia. Norman Borlaug, Nobel Prize winner, "In the next 25 years, farmers in Asia must increase their yields by 50-75%."

Combating Hunger

These benefits are particularly important considering about one billion people around the globe cannot count on eating every day, the World Bank. For example the populations of India and China are forecast to grow to 1.6 and 1.5 billion, respectively by 2050. The Philippine population is set to almost double over next 25 years. Simply put, the use of agricultural biotechnology can result in more food being produced on less land to feed a rapidly growing global population.

Approved Food Biotechnology is Safe

Opponents of biotechnology question its biosafety. However, many of the criticisms are not supported by evidence or fact. Government agencies involved in health, science, technology, agriculture and trade are committed to ensuring the safe application of biotechnology. A number of other health and food organisations also support the scientific basis of biotechnology. These include UN FAO, WHO, OECD, The Third World Academy of Sciences,the Institute of Food Technologists and the Australian Institute of Food Science and Technology. By May 2001, more than 3,000 leading scientists, including two Nobel Prize winners, had signed a declaration endorsing gene technology as a safe, environmentally-friendly and useful tool to help feed the developing world.

Labelling Policy

In Asia, national food standards have been developed in all countries to ensure the safety of the food supply. Standards for the labelling of foods produced through biotechnology are currently under discussion in most Asian countries and also by the United Nations agencies responsible for global food guidelines. In Japan, food ingredients produced through biotechnology must be labelled if they make up the first five ingredients by weight. Refined products such as oils and sugars in which no protein or recombinant DNA can be detected are excluded from labelling requirements. Korea has introduced mandatory labelling for all genetically modified foods with more than 1% genetically modified content. The debate around the labelling of genetically modified foods stems from the consumers "right to know" rather than a safety issue. All foods that are allowed to be sold in the market are regulated by the country's food regulations and are considered wholesome and safe to eat.

What does the Future Hold

The future for food made using agricultural biotechnology looks promising. New foods, nutritionally enhanced to protect and promote good health, are in development. Advances in biotechnology can help further reduce toxins in plants and help to detect plant contaminants more efficiently, thus improving the safety of our food supply. Fruits and vegetables may stay fresher for longer periods of time. Current advances suggest it may be possible to eliminate a number of common allergens from many foods. The available global food supply could be increased to support the rapidly growing world population.

Index